Leaving No Stone Unturned

Pathways in Organometallic Chemistry

F. Gordon A. Stone

PROFILES, PATHWAYS, AND DREAMS
Autobiographies of Eminent Chemists

Jeffrey I. Seeman, Series Editor

American Chemical Society, Washington, DC 1993

CHem

Library of Congress Cataloging-in-Publication Data

Stone, F. Gordon A. (Francis Gordon Albert), 1925–
  Leaving no stone unturned: pathways in organometallic
chemistry / F. Gordon A. Stone.

  p. cm.—(Profiles, pathways, and dreams, ISSN 1047–8329)

  Includes bibliographical references and index.

  ISBN 0–8412–1826–9 (cloth).—ISBN 0–8412–1827–7 (pbk.)

  1. Stone, F. Gordon A. (Francis Gordon Albert),
1925–   . 2. Chemists—Great   Britain—Biography.
3. Organometallic chemistry—History—20th century.

  I. Title. II. Series.

QD22.S76A3
540′.92—dc20
[B]   93–28909
                                                          CIP

Jeffrey I. Seeman, Series Editor

The paper used in this publication meets the minimum requirements of American National Standard for Information Sciences—Permanence of Paper for Printed Library Materials, ANSI Z39.48–1984.

∞

1993 Advisory Board

Foreword

In 1986, the ACS Books Department accepted for publication a collection of autobiographies of organic chemists, to be published in a single volume. However, the authors were much more prolific than the project's editor, Jeffrey I. Seeman, had anticipated, and under his guidance and encouragement, the project took on a life of its own. The original volume evolved into 22 volumes, and the first volume of Profiles, Pathways, and Dreams: Autobiographies of Eminent Chemists was published in 1990. Unlike the original volume, the series was structured to include chemical scientists in all specialties, not just organic chemistry. Our hope is that those who know the authors will be confirmed in their admiration for them, and that those who do not know them will find these eminent scientists a source of inspiration and encouragement, not only in any scientific endeavors, but also in life.

M. Joan Comstock
Head, Books Department
American Chemical Society

Contributors

We thank the following corporations and Herchel Smith for their generous financial support of the series Profiles, Pathways, and Dreams.

Akzo nv

Bachem Inc.

DuPont

Duphar B.V.

Eisai Co., Ltd.

Fujisawa Pharmaceutical Co., Ltd.

Hoechst Celanese Corporation

Imperial Chemical Industries PLC

Kao Corporation

Mitsui Petrochemical Industries, Ltd.

The NutraSweet Company

Organon International B.V.

Pergamon Press PLC

Pfizer Inc.

Philip Morris

Quest International

Sandoz Pharmaceuticals Corporation

Sankyo Company, Ltd.

Schering–Plough Corporation

Shionogi Research Laboratories, Shionogi & Co., Ltd.

Herchel Smith

Suntory Institute for Bioorganic Research

Takasago International Corporation

Takeda Chemical Industries, Ltd.

Unilever Research U.S., Inc.

Profiles, Pathways, and Dreams

Titles in This Series

About the Editor

JEFFREY I. SEEMAN received his B.S. with high honors in 1967 from the Stevens Institute of Technology in Hoboken, New Jersey, and his Ph.D. in organic chemistry in 1971 from the University of California, Berkeley. Following a two-year staff fellowship at the Laboratory of Chemical Physics of the National Institutes of Health in Bethesda, Maryland, he joined the Philip Morris Research Center in Richmond, Virginia. In 1983–1984, he enjoyed a sabbatical year at the Dyson Perrins Laboratory in Oxford, England, and claims to have visited more than 90% of the castles in England, Wales, and Scotland.

Seeman's 90 published papers include research and patents in the areas of photochemistry, nicotine and tobacco alkaloid chemistry and synthesis, conformational analysis, pyrolysis chemistry, organotransition metal chemistry, the use of cyclodextrins for chiral recognition, and structure–activity relationships in olfaction. He was a plenary lecturer at the Eighth IUPAC Conference on Physical Organic Chemistry held in Tokyo in 1986 and has been an invited lecturer at numerous scientific meetings and universities. Currently, Seeman serves on the Petroleum Research Fund Advisory Board. He continues to count Nero Wolfe and Archie Goodwin among his best friends.

Contents

Photographs

Preface

"HOW DID YOU GET THE IDEA—and the good fortune—to convince 22 world-famous chemists to write their autobiographies?" This question has been asked of me, in these or similar words, frequently over the past several years. I hope to explain in this preface how the project came about, how the contributors were chosen, what the editorial ground rules were, what was the editorial context in which these scientists wrote their stories, and the answers to related issues. Furthermore, several authors specifically requested that the project's boundary conditions be known.

As I was preparing an article[1] for *Chemical Reviews* on the Curtin–Hammett principle, I became interested in the people who did the work and the human side of the scientific developments. I am a chemist, and I also have a deep appreciation of history, especially in the sense of individual accomplishments. Readers' responses to the historical section of that review encouraged me to take an active interest in the history of chemistry. The concept for Profiles, Pathways, and Dreams resulted from that interest.

My goal for Profiles was to document the development of modern organic chemistry by having individual chemists discuss their roles in this development. Authors were not chosen to represent my choice of the world's "best" organic chemists, as one might choose the "baseball all-star team of the century". Such an attempt would be foolish: Even the selection committees for the Nobel prizes do not make their decisions on such a premise.

The selection criteria were numerous. Each individual had to have made seminal contributions to organic chemistry over a multidecade career. (The average age of the authors is over 70!) Profiles would represent scientists born and professionally productive in different countries. (Chemistry in 13 countries is detailed.) Taken together, these individuals were to have conducted research in nearly all sub-specialties of organic chemistry. Invitations to contribute were based on solicited advice and on recommendations of chemists from five continents, including nearly all of the contributors. The final assemblage was selected entirely and exclusively by me. Not all who were invited chose to participate, and not all who should have been invited could be asked.

A very detailed four-page document was sent to the contributors, in which they were informed that the objectives of the series were

1. to delineate the overall scientific development of organic chemistry during the past 30–40 years, a period during which this field has dramatically changed and matured;

2. to describe the development of specific areas of organic chemistry; to highlight the crucial discoveries and to examine the impact they have had on the continuing development in the field;

3. to focus attention on the research of some of the seminal contributors to organic chemistry; to indicate how their research programs progressed over a 20–40-year period; and

4. to provide a documented source for individuals interested in the hows and whys of the development of modern organic chemistry.

One noted scientist explained his refusal to contribute a volume by saying, in part, that "it is extraordinarily difficult to write in good taste about oneself. Only if one can manage a humorous and light touch does it come off well. Naturally, I would like to place my work in what I consider its true scientific perspective, but . . ."

Each autobiography reflects the author's science, his lifestyle, and the style of his research. Naturally, the volumes are not uniform, although each author attempted to follow the guidelines. "To write in good taste" was not an objective of the series. On the contrary, the authors were specifically requested not to write a review article of their field, but to detail their own research accomplishments. To the extent that this instruction was followed and the result is not "in good taste", then these are criticisms that I, as editor, must bear, not the writer.

As in any project, I have a few regrets. It is truly sad that Egbert Havinga and Herman Mark, who each wrote a volume, and David Ginsburg, who translated another, died during the course of this project. There have been many rewards, some of which are documented in my personal account of this project, entitled "Extracting the Essence: Adventures of an Editor" published in *CHEMTECH*.[2]

Acknowledgments

I join the entire scientific community in offering each author unbounded thanks. I thank their families and their secretaries for their contributions. Furthermore, I thank numerous chemists for reading and reviewing the autobiographies, for lending photographs, for sharing information, and for providing each of the authors and me the encouragement to proceed in a project that was far more costly in time and energy than any of us had anticipated.

I thank my employer, Philip Morris USA, and J. Charles, R. N. Ferguson, K. Houghton, and W. F. Kuhn, for without their support Profiles, Pathways, and Dreams could not have been. I thank ACS Books, and in particular, Robin Giroux (production manager), Janet Dodd (senior editor), Joan Comstock (department head), and their staff for their hard work, dedication, and support. Each reader no doubt joins me in thanking 24 corporations and Herchel Smith for financial support for the project.

I thank my children, Jonathan and Brooke, for their patience and understanding; remarkably, I have been working on Profiles for more than half of their lives—probably the only half that they can remember! Finally, I again thank all those mentioned and especially my family, friends, colleagues, and the 22 authors for allowing me to share this experience with them.

JEFFREY I. SEEMAN
Philip Morris Research Center
Richmond, VA 23234

April 7, 1992

[1] Seeman, J. I. *Chem. Rev.* **1983**, *83*, 83–134.
[2] Seeman, J. I. *CHEMTECH* **1990**, *20*(2), 86–90.

Editor's Note

THE EMINENT CHEMIST HENRY GILMAN, in a congratulatory letter to F. Gordon A. Stone on the occasion of the latter's 65th birthday, wrote, "You are a chemist's chemist, and as such will long serve as a classical reminder of innovative and truly profound research." Gilman also noted, "Your delightful sense of humor is penetrating and most friendly." That insightful description of Stone's humor could apply equally as well to other facets of his personality, with "penetrating" the operative term. All those who know Gordon Stone can certainly attest to his intensity and dedication to his career; he is perceived as a scientist who allows no distractions in his devotion to his profession. "I have been lucky to do what I liked," Stone has summed up his accomplishments modestly.

As a result of his penetrating insight, Stone has received many honors. He is a Fellow of the Royal Society, which in 1989 bestowed upon him its prestigious Davy Medal. He has received several awards from the Royal Society of Chemistry, and the American Chemical Society has recognized him for his research in inorganic chemistry. The United Kingdom University Grants Committee appointed him chair of their chemical review committee and charged him with the responsibility of reviewing techniques for teaching and research in chemistry at British universities. The results of that study, published in 1988 as *University Chemistry: The Way Forward*, became known as "the Stone report." In it are proposals for the "general principles for the future of teaching and research in chemistry" in Great Britain, as he explained in the preface to the report.

And then Gordon Stone entered my life and changed many of my perceptions of my duties as an editor with this series, *Profiles, Pathways, and Dreams*. He submitted what we all thought at that time was his final manuscript in April 1987. However, the project changed in scope and size from what was to be one chapter in a volume of 22, to a 22-volume series. He was asked to, and did in fact, enlarge his manuscript and broaden its scope considerably in his insightful and penetrating style. Several years and many pages of correspondence passed between us until, in response to my penultimate request for additional information, Gordon responded by sending me copies of some of my correspondence to him, in chronological order and heavily scored and annotated. "Worth scanning" he had noted on the first page, letting me know with "penetrating and most friendly humor" that he had indeed responded to various requests, considered my comments and concerns, and that he felt the task had been completed to *his* satisfaction.

My actions and style as series editor came under self-scrutiny. In scanning the correspondence that was the history of our mutual task, I took note of his marginal notes and my many wishful promises of "this is the last time . . ." that had become an unintentional but recurring theme. It is vital, I concluded, to be kept honest, and Gordon Stone is the man to do just that. A comment Stone made about himself seemed most apropos to the moment: "I tend to think of myself as an irascible fellow on occasion, but open to reason, and I think I have a good sense of humor. My wife says that you have not sufficiently conveyed that I am an impatient devil and somewhat aggressive, but I think these feelings are rather common among many researchers."

It was with bemusement that I reexamined my relationship with Gordon: the editor of the *Profiles* series interacting with the experienced editor of many volumes. And I comforted myself with the memory of a comment from one of his friends that "Gordon has mellowed considerably over the years. He is much more comfortable with himself." My experiences with Gordon Stone have been stimulating and rigorous in an atmosphere of cordiality. At times, he has held me to the task rather than the reverse!

Stone is as reticent about the personal aspects of his life as he is open about his science. "This is due to my British temperament," he explained during one of my many unsuccessful attempts to get him to include more of his personal history in this volume. Yet, upon reading the draft for this article, he suggested I contact a few of his students to ask *them* "what they learned from me besides chemistry." Indeed, Stone was able to identify the sources of some of my unnamed quotations from students and long-time friends and colleagues. "I observed the use of Herb's [Kaesz] current buzzwords: 'Gordon has mellowed considerably over the years.' He says this every time we meet!"

Actually, it was a visit to his office and laboratory in Bristol some years back that led me to contact his students. His office had two doors, one to an anteroom housing his two secretaries, through which all visitors must pass, and the second into his students' laboratory. Over that second door, on the laboratory side, were several lights—red meant wait, ("but I never press it"); green meant come in. "There is no barrier to my students," he explained. "My students never need to make an appointment. I am like a pastor who looks after his parishioners." The depth of his concern for his students triggered my desire to learn more about his role as a teacher. One student praised Stone's "demanding but very nurturing environment. He has a way of challenging a student. He always has time for us, early or late, and he walks us through a thought process. His questions prompt our thinking. He listens to us talk our way to the answers."

"Prof is just a treasure! He has an incredible sense of humor. He cares for the whole of an entity, not just a part. He demanded that we first understand what was going on in our particular experiment and then make hypotheses on a more general scope. Specific events led to general concepts. This approach was applied to his students, as well. He cares for the whole of the individual, just as he cares for the whole of chemistry."

"It's very hard to work in my group and not get a publication as a result," Stone observed. "Many co-workers need 'bread-and-butter' problems for them to gain confidence in their experimental work, and it is important careerwise for them to have some published work to show for their efforts. Of my 600 or so primary journal articles, perhaps fewer than 50 have strongly influenced the way people think about chemistry. Chemistry is a team effort by many individuals slowly constructing a large edifice of knowledge. Most of us can only hope to place a few useful building blocks into the scheme of things."

Judy Stone has shared her husband's continuing interest in and concern for his students. As one former pupil explains fondly, "You can ask Judy about any student, even one who studied with Gordon 20 years ago, and she will know his or her spouse's name, where they are, the number of children they have, and their ages! It is almost a parental atmosphere."

Stone is a complex individual; his demanding nature is balanced by his humor, his praise, and his loyalty. One student laughingly recalled that prior to the annual picnic at his home in Bristol, Stone had mowed the term "W≡C" into his lawn as a focus for team celebration. I know from personal experience that Stone's praise has influenced me and increased my own motivation. (And his criticism is just as effective!)

He is also a selfless individual, involving himself in various pursuits not necessarily having anything to do with his own science. He has been active—indeed, proactive—in promoting chemistry in Britain. In addition to his work on the aforementioned Stone report, he has served on a number of committees for the Royal Society, the Royal Society of Chemistry (RSC), and the Science and Engineering Research Council (SERC). He served as vice-president of the Royal Society in 1987–1988 and president of the Dalton Division of the RSC in 1981–1983. I know him to be one of the most organized, thorough, and devoted professionals; he meets deadlines and never shirks from commitments. Any committee that includes Stone will accomplish its goals and make significant contributions.

Stone's organizational skills were vividly and visually, though certainly unintentionally, communicated to me on my visit with him in Bristol. Thirty or more mail trays were carefully placed about his office. These were the homes of manuscripts in various stages of production. (The tray for this volume indeed must have been quite full, without even counting our letters and faxes!) A sign over his door read, "I may have my faults, but being wrong is not one of them." Directly above this sign was a picture of a large elephant crashing through the African bush. Gordon chuckled, when he noticed me gazing at it. "It is a good statement for the head of the department," he explained. He did not take the sign with him when he left Bristol, however. Upon his reading of this essay, Gordon wrote, "It is great that you remember seeing it [the sign and the picture], because I had forgotten its existence. The sign was given to me by Judith Howard. We had many friendly arguments that I think—hope?—typified my relationship with her and other colleagues during my 27 years at Bristol."

According to a colleague, Stone has a "green thumb" for synthesis. He has a love for the complexity of molecules and the talent to make them. Stone's name came up in a recent telephone conversation I had with Roald Hoffmann, the well-known Nobel laureate. The note he had written Stone to congratulate him on receiving an honorary degree at Exeter, reproduced here, is an indication of the synergy between experiment and theory, and of the mutual respect between the experimentalist and the theoretician.

Gordon Stone is the chemist's chemist, the experimentalist's experimenter. He is as solid and as visible as the compounds he and his group have been making for decades.

JEFFREY I. SEEMAN
Philip Morris Research Center
Richmond, VA 23234

June 4, 1993

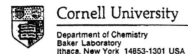

Cornell University

Department of Chemistry
Baker Laboratory
Ithaca, New York 14853-1301 USA

July 17, 1992

Congratulations, Gordon, on your newest
degree!

Your continued enthusiasm for inorganic
chemistry, your magic skill at making
molecules (and organizing others to help
you do so), your obvious delight at
the bounty of nature, has been an
inspiration to your colleagues.

What has always impressed me is
how you could know ahead of time what
the theoreticians would later come up
with! So while I benefitted from your
popularization of the isolobal analogy
as a synthetic tool, I know full well
that you didn't need that analogy at
all, that you know deeply and surely,
what it took me and others, in our
roundabout way, years to find.

Stay well, friend

Roald (Hoffmann)

Roald Hoffmann, in his unique, lyrical style, sent his congratulations to Gordon Stone, who was awarded an honorary degree from Stone's alma mater, Exeter. Hoffmann composes much of his correspondence in his own hand, as exemplified here. His holographs to me are especially treasured.

xxiii

This book is dedicated to Judith Stone

Leaving No Stone Unturned

I WAS FORTUNATE TO COMMENCE MY RESEARCHES in organometallic chemistry at a time when the subject was just beginning to attract strong interest in both academic and industrial laboratories, and when the number of investigators was still relatively small. Most experiments yielded exciting and often unexpected results, and new types of organometal complex were reported in almost every new issue of the relevant journals. One learned quickly to visit the library on a daily basis in order not to miss a new discovery. It is nostalgic now to recall that funding agencies such as the National Science Foundation and the Air Force Office of Scientific Research were seeking chemists to support. One never seemed to hear of a rejected grant proposal, since it was the "Sputnik era"! The confidence of the funding agencies in supporting research in organometallic chemistry as a worthwhile endeavor, coupled with the ability of these sponsors to provide the necessary money, was amply rewarded by the extraordinary progress that was made over a relatively short period. Indeed, organometallic chemistry has grown to such an extent that it has become a major domain of our science.[1]

Organometallic chemistry is generally defined as the study of compounds containing carbon—metal bonds, and this area of endeavor has enormous breadth because of the host of different ways in which metals are able to modify the reactivity of organic groups through coordination, or in which the metals are themselves influenced by substrate molecules to adopt a variety of bonding modes and reactivity patterns. Even today, some 40 years after the elucidation of the structure of ferrocene, unexpected chemical behavior is frequently encountered in the study of organometallic compounds. Our knowledge of chemical bonding, structure, and mechanisms continues to increase as activity in this area is maintained at a high level. Moreover, organometallic compounds are now used to obtain useful products ranging from plastics in common everyday use, through bulk organic chemicals such as acetic acid, to drugs such as L-dopa. Metal-vapor deposition by decomposing suitable organometallic precursors is beginning to become useful in the electronics industry. Thus a collection of autobiographies by active practitioners of organic chemistry would be incomplete without a contribution from an organometallic chemist, and this is my main justification for writing this book.

Early Years

Until the age of 15, I attended several state and private schools in the south of England, but I remained at each school for only a relatively short period because my father's place of work was changed. He was employed in the civil service, and as he was transferred between different office locations we moved our home. In 1940 we were living in Margate in East Kent, and in that year I was sent to Exeter School in Exeter, a city in the southwest of the country where I had been born in 1925, the only child of my parents. During the 1930s times were very hard financially for my parents, as they were for many others in that decade. As a government servant my father was never unemployed, but he and others like him had their salaries reduced periodically. Nevertheless, I was well cared for, and my education was given a very high priority by my parents, enabling me to attend good schools. Moreover, they ensured that I met all my homework assignments before I carried on other activities. My father assisted me greatly in introducing me to the elementary principles of mathematics.

My return to Exeter in 1940 was a consequence of East Kent during World War II becoming an area subjected to repeated bombing and even long-range shelling, and all the schools were closed or evacuated to places in the north or west of the country. I went to live with my maternal grandparents in Exeter during the school terms, but I returned to Margate during vacations, thereby having the opportunity to see some of the air

With a friend, circa 1935.

activity by both sides over East Kent and the Channel, which was exciting for a school boy. By chance I was on vacation in Margate when Exeter was heavily bombed in May 1942. There were some distinct advantages to spending school vacations in Margate in that period. That part of Kent was restricted for entry to those who were permanent residents because of its proximity to the French coast. Much of the permanent population had retired inland, and consequently the local golf course at the North Foreland was virtually deserted, so my friends and

At school on a sports day, circa 1938.

My parents, circa 1954.

My maternal grandmother, Frances Ellen Coles (circa 1935), in whose home I stayed during my school years.

I spent many a day there uninhibited by the presence of other golfers. Unfortunately, other calls on my time led me to abandon the game after I had left university.

I owe my grandparents a great debt for providing me with a good environment for study when I was in Exeter. There was always a quiet room available for homework after school, and no shortage of food at a time of tight rationing during the war. Very likely my grandparents allocated some of their rations to me. My grandfather had retired from a business of furniture-making and related carpentry activities when I lived in his house. I retain in my home a bed of elegant design he made for me. Although he had probably received no formal academic education, he was very well read, with a particularly good knowledge of history. Consequently, many books of an historical and biographical nature were always on hand, and he often reviewed their contents with me. This atmosphere generated in me a lifelong interest in reading as a means of relaxation, particularly if the books are biographical or nonfictional in nature. Moreover, sharing reading interests with my grandfather, and also with my father, very probably helped me develop a retentive memory, and also ingrained in me the ability to assimilate information rapidly—valuable attributes for a chemist, for whom success in research is assisted by keeping up to date with the mass of current literature.

Exeter School, like many schools in Britain in that era, was operating under considerable difficulties, including loss of teaching staff for war service. Consequently, academic life was somewhat disrupted. My parents thought that I should take up accountancy as a career, but I developed an enthusiasm for chemistry while at school, as probably happens with most people who take up the subject as a career. Chemistry was my best subject, and I was excited by the synthesis, analysis, and colors of the simple compounds we studied in the Exeter School laboratory. Unfortunately I do not recall the names of the many teachers who taught us chemistry and physics in the period 1940–1944, but I do remember the name of one singularly effective school teacher, a Mr. Bernard Foster, and he perhaps more than any other person generated in me a great liking for the subject. Some of the science teaching was conducted in a stimulating but ad hoc manner by members of staff "on loan" from the local University College, which at that time was an external branch of the University of London. After the Second World War the college grew in size and became independent as the University of Exeter, with its own charter. I was particularly pleased when in 1992 I was awarded an honorary degree of doctor of science by this university in the city where I was born. As an accountant I would have made more money, but I know I would not have enjoyed myself so much, and very probably I would not have had so many opportunities to travel and meet interesting people.

I planned to attend London University after leaving school, studying chemistry at either King's College or at the Imperial College of Science and Technology. However, very fortunately as it turned out, I was rejected by these two institutions because the "grades" I had obtained in my final examinations at school were not deemed satisfactory. Indeed I had failed the examination in my best subject, chemistry, although I obtained good passing grades in physics and mathematics. Although they were not welcome at the time, I later came to treasure my rejection letters, which were addressed in those far gone paternalistic days to my father: one is reproduced on this page. Some 20 years later I was considered sufficiently qualified to become an external examiner for chemistry at Imperial College, and during my time at Bristol University I was appointed to serve on several advisory committees for the various colleges of London University, including some for the appointment of candidates to chairs.

My rejection by London led to my spending a year at the Medway Technical College at Gillingham in Kent, with the object of improving my knowledge of chemistry and taking examinations for direct entry into London University at the end of 1945. However, while recovering from the denial of entry to King's and Imperial Colleges of London, I decided to apply to Cambridge University on the basis of the very same examination grades that had been deemed unsatisfactory by those in charge of student admissions at the two London colleges. Although strongly encouraged by my parents to exploit all available educational opportunities, the decision to apply at Cambridge was my own. Although my paper qualifications in chemistry might seem weak, at least I had attained the necessary standard in Latin, a prerequisite at that time for entry to Cambridge.

UNIVERSITY OF LONDON KING'S COLLEGE.

FACULTY OF NATURAL SCIENCE.

TEMPLE BAR 5651.   STRAND. W.C.2.

29th May 1945.

Dear Mr. Stone,

F.G.A. Stone.

As I suggested in my letter of 3rd November 1944, there has been keen competition for places in the College for next session and particularly for places in Honours Chemistry. The Professor of Chemistry has drawn up his first list and placed ten students on a reserve list and he has decided that your son's name cannot be placed in either of these lists. As I do not think even the ten reserves will obtain places in the Chemistry School I hope you will have found a place for your son at another College.

Yours sincerely,

James Henderson.

S.C. Stone Esq.,   Sub-Dean of the Faculty of Science.
Dunkery Beacon,
Northumberland Avenue,
Margate.

No one in our family had attended a university, and there seemed to be a generally held view that London University was the place to go to work and to study science, and that Cambridge and Oxford were the universities to attend if you were a gentleman, had sufficient money, and wanted to enjoy yourself. However, by the end of the Second World War it was no longer true that a "Brideshead" spirit was a major contributor to student life at "Oxbridge". Nevertheless, in Great Britain this was an era when only about one young person in 20 proceeded into any form of higher education after leaving school, as compared with one in four by the 1990s. Although having little money, I applied for entry to three Cambridge colleges simultaneously and was accepted by all three on the basis of a recommendation from my headmaster. He must have written a good letter.

Because the first reply to my applications arrived in the mail from Christ's College, I accepted this offer, and I entered Christ's as an undergraduate in October 1945. Subsequently, I never received a poor examination result and gained first class honors and prizes in both parts of the Natural Sciences Tripos. The lesson I learned from my rejection by London University was to have faith in one's own ability and not to be easily rebuffed by the actions of others.

For those unfamiliar with the Cambridge system, the Tripos examinations are those taken at the end of the second and the third year of study in most subjects, the degree course being 3 years. My very satisfactory performance in the examinations probably reflects the high quality of the undergraduate teaching in inorganic and organic chemistry at Cambridge at that time and hard work to assimilate what was taught. Unfortunately, much of the instruction in physical chemistry seemed very incoherent, although I remember Fred (now Lord) Dainton's classes on thermodynamics as models of clarity. There was also some stimulating laboratory class supervision from two young assistants, George Porter and Norman Sheppard, who were later to have very distinguished careers. Apart from these persons, if you survived in physical chemistry you were largely self taught, but perhaps this method of learning has its merits in developing initiative. The students of today seem to demand the distribution of lecture synopses from their instructors and are unwilling to read widely about the subject.

In 1945–1948 the life of an undergraduate at Cambridge was not easy. Even though World War II had ended, food rationing continued, and for some items was even more restrictive than during the war. Fuel for heating, usually by open coal fires, was in short supply. The student body had rapidly expanded as men returned to study after war service. Because of this, and for other reasons, the colleges had insufficient accommodation. Consequently, most students lived in rented rooms in Cambridge. Much of this accommodation, located in houses of the early Victorian era, was very primitive even by the standards of that day and fortunately has long disappeared. However, responding to the calls of nature by walking to an outside toilet at the end of a garden on a night when a hard frost gripped the Fen country was perhaps a character-building exercise for which in retrospect one should be grateful. The plumbing situation was hardly improved when for a time I occupied rooms in college purported to have been used by John Milton in the early seventeenth century. Nevertheless, I recall my period as an undergraduate with pleasure. Many extracurricular activities were on offer, both formal and informal, and like most of my fellow students, not all my time was devoted to study.

In spite of some recruiting pressures from Lord Todd's associates to join his large and already internationally renowned research group in natural product chemistry, which included at that time such future stars as Arthur Birch and H. G. Khorana, I decided to undertake graduate studies in inorganic chemistry. I had been particularly intrigued in my final year as an undergraduate by the interesting subject matter of the lectures delivered by Professor Harry Eméleus. He had come to Cambridge from Imperial College, after working in the United States on the Manhattan Project, and he had already developed a strong research group in fluorine chemistry. Although I became one of Harry Eméleus's graduate students in 1948, I did not work with him on fluorine compounds, but on boron hydrides. This came about because the British Admiralty had asked Professor Eméleus to initiate some research in his laboratory on boron hydrides and their derivatives, as a consequence of an interest that the U.S. Navy had developed in these compounds. During World War II the United States Office of Naval Research (ONR) had sponsored, under Professor H. (Herman) I. Schlesinger at the University of Chicago, a project to seek solid compounds

Research workers and the academic staff in organic, inorganic, and theoretical chemistry at Cambridge University in 1950. I am standing fifth from the right in the fourth row from the front, with future Nobel laureate H. G. Khorana on my left. A. J. Birch, a contributor to Profiles, Pathways, and Dreams is standing third from the left in the second row. The laboratory was under the direction of Professor Todd, later Lord Todd, who is seated at the center front, flanked on his left by the distinguished theoretical chemist Sir Lennard Jones and on his right by Professor Harry Emeléus, who was the first professor of inorganic chemistry at Cambridge. The absence of any physical chemists in the photograph was due to the administrative division of chemistry at Cambridge at that time; the physical chemistry laboratory was under R. G. W. Norrish, and all other sectors of the subject were under A. R. Todd. Both Todd and Norrish were later to receive Nobel Prizes for chemistry, the latter in association with George Porter.

As a graduate student at Cambridge University in 1949. The first research on the boranes in Great Britain; I used a vacuum line constructed from soda glass, since Pyrex was unavailable.

able to act as sources of hydrogen upon hydrolysis to avoid the necessity of transporting this gas in heavy cylinders in submarines, or on land in certain circumstances. Shortly before the entry of the United States into the Second World War, the Chicago School had discovered the first borohydrides: $LiBH_4$, $Be(BH_4)_2$, and $Al(BH_4)_3$. These compounds were known to hydrolyze readily, but for one reason or another were unsuitable for the desired purpose; for example, $Al(BH_4)_3$ is a liquid that explodes on contact with water and is therefore far from a suitable source of "portable" hydrogen. The Chicago workers, however, subsequently discovered $NaBH_4$ and $LiAlH_4$, which were more useful, particularly the former. Later they were to characterize sublimable $U(BH_4)_4$, the relative volatility of which, compared with other uranium species, was of interest very briefly to those involved in the Manhattan Project who were concerned with separating uranium isotopes by diffusion. However, the technology of handling fluorine and volatile fluorides was developed to a level whereby it was possible to use the more volatile UF_6 for this purpose. Nevertheless, the preparation of the important reagents $NaBH_4$ and $LiAlH_4$ by Herman Schlesinger's group during the 1940s is now recognized as having been a very major contribution to chemistry, because of their subsequent proven value in synthesis.

When I began research, the results of the studies at Chicago on the borohydrides and related species had not appeared in any of the open scientific literature. However, the ONR reports, in which the work was described, were available to us via scientists at the British Admiralty, so that one knew of the existence of $LiAlH_4$ and that treatment of this complex hydride with BCl_3 in ether gave B_2H_6 in good yield. To begin work in the United Kingdom in this area it was necessary, as a new research student, to prepare both BCl_3 and $LiAlH_4$ so that these compounds could subsequently be used to generate B_2H_6. A Stock-type high-vacuum system with greaseless valves also had to be constructed from soda glass, since Pyrex glass was not available. I recall with amusement various activities in the laboratory that would in no way satisfy the health and safety regulations of today. Liquid nitrogen was in very short supply, so liquid air, with its high oxygen content, was in general use for cooling purposes. For the record it must be stated that most of the various explosions that occurred in the Emeléus group at that time originated with the dozen or so students and postdoctoral assistants working with fluorine compounds, and not with the single person on the borane project. The only useful physical method for following reactions was infrared spectroscopy with a very primitive instrument by the standards of today. Not surprisingly, it took almost a year to generate a millimole of B_2H_6, probably the first sample of this compound ever to be made in the United Kingdom. One started from elemental boron to make BCl_3, and from $AlCl_3$ for the $LiAlH_4$ synthesis. In my Ph.D. examination one of the examiners told me that I should not expect to be able to continue research on the boranes because of the high cost of the mercury required for use in the Stock-type high-vacuum systems. At that time the British welfare state had not gained full momentum, so students paid a deposit against breakage of glassware or equipment. In subsequent times breakage deposits disappeared. However, I paid little heed to the advice about the high cost of mercury.

During 1951, the last year of my graduate work, it became necessary, if perhaps unwelcome, to address the problem of what I was going to do after completing my Ph.D. Harry Emeléus's students seemed at that time in my inexperienced eyes to fall naturally into three categories: those clever enough to be kept on at Cambridge in some college post, those who went into industry, and those who crossed the Atlantic to gain postdoctoral experience in the New World. It was obvious that I did not fall into the first category, and the idea of spending a lifetime in industry, probably in the north of England,* did not appeal. Therefore, the decision as to what to do next was easy for me, and, having accomplished relatively little with my Ph.D. project, in 1951 I joined Anton Burg's group at the University of Southern California (USC) as a postdoctoral scholar under the Fulbright program. I thought I would be in the United States for one year. Little did I know that 10 years would elapse before I would return to work in Great Britain.

Southern California was a very welcome change after the austerity of postwar Britain. I remember shortly after my arrival going with Anton Burg across the USC Campus to the cafeteria in the sunshine, with clear views of the snowcapped mountains near San Bernadino on the horizon. Although Burg proceeded at what was his normal walking pace, I had to trot along to keep up with him. Anton had been a champion high jumper at university, and in his youth was purported to be able to do standing jumps over seated social groups. At the time I was at USC, some World War II barrack buildings of wooden construction were in use as laboratories, and there was only one exit from each floor. I was told that this had been of some concern during a fire safety inspection just before I arrived, and it seemed possible that the labs would be condemned. However, during the inspection, Anton rose to the challenge that the

---

* Most industrial research laboratories were located at that time in the Midlands or in the north of England. Coming from Devon, in the southwest of the country, my knowledge of anything north of a line through Oxford and Cambridge was minimal, except for an awareness that there would be even less sunshine in these nether regions.

department might lose valuable laboratory space by vaulting out of the window of a second-floor laboratory and landing perfectly upright and composed on the grass outside. Apparently the fire inspector was satisfied by this demonstration of procedure for evacuation of the building, because these very hot and uncomfortable laboratories remained in use for some years. This story will explain why as a young postdoctoral assistant it was necessary for me to move fast to keep up with Anton Burg.

There was, as I recall, only one freeway in the L.A. area at that time, running from downtown Los Angeles to Pasadena. Smog was noticeable in the Los Angeles basin on only a very few days in the fall of 1951. Anton, however, drew attention to this growing problem by riding his bicycle around the streets while wearing a World War II gas mask, a practice that gained him some publicity in the *Los Angeles Times*. Since Anton Burg did not drive a car his bicycle was his means of transportation.

The USC Chemistry Department was a very lively and friendly place to work, and there were several postdoctoral assistants from Europe, Australia, and New Zealand. The department chairman was Professor Robert Vold, a colloid chemist, and he, his wife Marjorie, and his colleagues, including Arthur Adamson, Sidney Benson, Norman Kharasch, Karol Mysels, and James Warf, were extraordinarily hospitable to all the postdoctoral visitors, and as a group we received frequent invitations to their various homes for dinners or parties.

Professor Burg had moved to USC as chairman of the chemistry department in 1940 from the Schlesinger group at Chicago, and by the time I reached Los Angeles borane chemistry had developed extensively. The major activity in Burg's laboratory when I arrived involved the synthesis of thermally and hydrolytically stable inorganic oligomers and polymers based on B—N, B—P, and B—As bonds, an area for which there was considerable ONR and other federal agency funding at that time. In Anton Burg's laboratory one learned by example from him the meaning of "motivation" and that success in chemistry is most likely to result from careful experimentation and especially hard work. It is a pleasure to acknowledge the stimulation I received in my 2-year association with Anton, although no one could rise to the standards he set for hours spent in the laboratory. I do not remember his ever being away on vacation, apart from Christmas day. I was, therefore, not surprised to note that at the

Professor Anton B. Burg (University of Southern California), under whom I carried out postdoctoral work in the period 1951–1953. Anton is standing in front of a high-vacuum system with mercury float valves of the kind which originated with Alfred Stock for the manipulation of boranes and silanes. Fortunately, the advent of both greaseless stopcocks and lubricants inert to most reactive reagents has eliminated the need to employ mercury float valves in vacuum systems.

age of 86 he recently published a paper in *Inorganic Chemistry* based solely upon experimental work done with his own hands. Anton provided his group with superb training in the manipulation of compounds in the absence of air, which in my case proved invaluable when later I took up research in organometallic chemistry and could teach my students to use high-vacuum systems or to conduct experiments under an atmosphere of nitrogen.

Following a meeting I had with Professor E. G. (Gene) Rochow at an ACS meeting held in Chicago in September 1953, I moved to Harvard in February 1954 to join his group as a postdoctoral assistant. However, after the ACS meeting I returned briefly to England, where I took the opportunity of discussing my future career prospects with Harry Emeléus. Since there seemed at that time to be no opportunities in academia in the United Kingdom, he suggested that I apply for a vacancy on the teaching staff of the chemistry department of the University of Singapore. The last rays of the sun were setting on the British empire at that time, and it was still just possible to dispatch doubtful characters to one or other of the colonies. However, I was lucky to have in my pocket the offer from Gene Rochow to join his group as a postdoctoral fellow in the older "colony" of Massachusetts.

The journey back to the United States was very fortunate for me in that while traveling on the *Queen Mary*, now resting at Long Beach, California, I met Miss Judith Hislop of Sydney, Australia, who subsequently became my wife. Shipboard journeys of an earlier era presented better opportunities to develop relationships than when crossing the Atlantic by the jet planes of today. It could be said that we were thrown together by the fact that the voyage occurred in February, under very stormy

Professor H. J. Emeléus, with me on his right and Professor W. (William) A. G. Graham on his left, at the Third Chemical Congress of North America held in Toronto, June 1988. At this Congress Harry Emeléus received an ACS award for his distinguished contributions to fluorine chemistry.

conditions, so that most of the other passengers were not seen from the day of departure from Southampton until we approached the docks in New York. As my academic career proceeded, my wife's great talent as a hostess became well known to numerous students and visitors. Moreover, she devoted much of her time and energy to smoothing the pathway for the wives of new postdoctoral co-workers upon their arrival, many of whom were found accommodation, even in our own home for short periods if necessary. Her activities ensured that new colleagues quickly adjusted to unfamiliar surroundings in a manner that would not have happened if I had been solely responsible for their affairs. They were thus able to begin to work effectively in my laboratory very quickly after arrival. Also, years later, when at Bristol I became head of a large department, she made sure we were in reasonably frequent contact with junior staff members and their wives, a feature that helped to smooth working relationships. The patience, good humor, and resilience shown by my wife has assisted my career in many ways.

Although I joined Professor Rochow's group as a postdoctoral assistant, I became an instructor within a few months, largely through his influence. At that time an instructor was the lowest academic rank at Harvard and carried a salary of $3000 per year. My responsibilities were to assist Gene Rochow in teaching the basic freshman course taken at that time by about 300 students. I very much enjoyed this experience for several reasons. Rochow was a first-class teacher, presenting well-organized lectures, and his wry humor was much appreciated by the students. He was not only a very lucid lecturer for students in an introductory course, but he also made a great effort to help during office hours those students who were encountering difficulties. As an aspiring academic one could learn a great deal from him about how to teach. Although my main responsibility lay in organizing the laboratory and the discussion sections, I had an opportunity to give some of the lectures when Gene Rochow was away from Harvard.

After I became an instructor I was able to start independent research, and my first graduate students, Bill Graham and Herb Kaesz, worked on aspects of boron chemistry, including a comparison of the Lewis acid properties of BH_3 and BF_3. I was interested to note recently that the results were mentioned in

Judith Stone, 1960.

the latest edition of the well-known inorganic text book by Al Cotton and Geoffrey Wilkinson, but without attribution, and very reasonably so because of the time that has elapsed since the work was done. In chemistry, because of its diverse nature and the multitude of workers contributing to the subject, memories soon fade as to which researchers were responsible for a particular piece of work in the first instance.

The Renaissance of Organometallic Chemistry

Before giving a personal account of the researches with which I have been associated, I will review the state of the art of organometallic chemistry at about the time my own work began. When I came to Harvard, Geoffrey Wilkinson* and his students, including at that time John Birmingham, Al Cotton, and Stan Piper, were developing metallacene chemistry at an astounding rate and discovering new classes of organotransition element complexes, including stable carbonyl-, alkyl-, and hydrido(cyclopentadienyl)metal compounds. All this exciting research, and the equally significant work on these compounds carried out at the same time by E. O. Fischer's school at the Technischen Universität in München, served to form a bridge between the classical domains of inorganic and organic chemistry and to remove the barrier between coordination chemistry and organometallic chemistry. As is well known, the latter field originated in the last century and towards the beginning of this century, with the seminal contributions of Zeise, Frankland, Mond, and Grignard being well known to most chemists. However, it was not until the isolation and characterization of ferrocene and other cyclopentadienylmetal "sandwich" compounds[2,3] that a spectacular increase in our knowledge of organometallic chemistry began. Although the growth of this field has been most

---

* For an entertaining account of early researches on the metallacenes, *see* G. Wilkinson, *J. Organomet. Chem.* **1975**, *100*, 273.

A collector's item: an informal photograph of Nobel laureates E. O. Fischer (right) and G. Wilkinson (left) taken at the Ettal Conference (1974), held to celebrate the 80th birthday of W. Hieber, generally regarded as the father of metal carbonyl chemistry.

marked by discoveries involving the d-block metals, and more recently those of the f-block, studies on the organoderivatives of the metals of groups 1, 2, 13, and 14 of the periodic table have also produced results of lasting importance. Prominent among these have been Herbert Brown's syntheses of organoboron compounds via hydroboration and Karl Ziegler's preparation of aluminum alkyls on a large scale and their involvement in alkene oligomerization and polymerization processes.

By the mid-1950s our knowledge of the chemistry of the cyclopentadienylmetal compounds and many of their derivatives had become relatively well developed as a result of the activities of the groups of E. O. Fischer[4] and G. Wilkinson.[5] In 1955 Fischer and Hafner[6] made a further major advance when they reported the synthesis of $[Cr(\eta^6\text{-}C_6H_6)_2]$. Remarkably, several related π-arene–chromium complexes had been prepared by Hein[7] more than 30 years previously, but the "sandwich" nature of these species was not recognized until a timely reinvestigation of these products by Zeiss and his co-workers[8] led to their correct identification. This study also led Zeiss and Tsutsui[9] to an independent synthesis of $[Cr(\eta^6\text{-}C_6H_6)_2]$. Interestingly, however, although the Zeiss–Tsutsui work eventually appeared in the *Journal of the American Chemical Society*, publication of the results was long delayed because the referees of the manuscript rejected the proposed π-arene structures on the grounds of insufficient evidence.* Although the idea of a metal atom hexahapto-coordinated to a benzene ring is a structural feature that we now accept without a second thought, the concept was evidently alien to some referees of that era. Since ferrocene was by then well known they should have had more imagination.

In parallel with the researches on cyclopentadienyl– and arene–metal complexes, there was a rapid upsurge of interest in the metal carbonyls, and these areas merged with the discovery of many compounds containing both carbonyl and hydrocarbon groups, for example, $[Fe_2(CO)_4(\eta^5\text{-}C_5H_5)_2]$ and $[Cr(CO)_3(\eta^6\text{-}C_6H_6)]$. Since 1927, Walter Hieber and his students[10] had developed metal carbonyl chemistry, discovering many important derivatives including the carbonyl hydrides—for example, $[FeH_2(CO)_4]$, the first transition element complexes known with metal–hydrogen bonds. Who would have imagined that what were then laboratory curiosities were actually the paradigm species of a class of molecule that was subsequently to form the basis of important industrial processes including hydroformylation? Hieber's contributions were remarkable in that many occurred prior to the availability of modern spectroscopic techniques or the use of X-ray diffraction on a routine basis. Not surprisingly some of the early formulations were incorrect. For example, Hieber's rhodium carbonyl "$[Rh_4(CO)_{11}]$" was later shown by X-ray diffraction to be $[Rh_6(CO)_{16}]$ in a study by L. F. Dahl et al.,[11] which revealed the first example of a molecular

---

* The subject matter of reference 9, based on the late Professor Tsutsui's Ph.D. thesis at Yale (1954), was reported at the September 1954 ACS Meeting (*see* footnote in reference 9). Hence the important results of Zeiss and Tsutsui were well documented prior to the publication of the synthesis of $[Cr(\eta^6\text{-}C_6H_6)_2]$ from benzene.[6]

Professor Walter Hieber at the conference on metal carbonyls held in his honor at Ettal (Bavaria) in 1974.

structure with carbonyl groups bridging three metal centers simultaneously. That carbonyl groups could bridge a single metal–metal bond had been known since the classic X-ray diffraction study on $[Fe_2(CO)_9]$ reported in 1939.[12,13]

An especially timely discovery in metal carbonyl chemistry was the synthesis of decacarbonyldimanganese by workers at Union Carbide, reported in 1954.[14] Subsequently an X-ray study of $[Mn_2(CO)_{10}]$ by Dahl and R. E. Rundle[15] was pivotal in establishing the first instance of a dimetal carbonyl species in which the two parts of the molecule are held together by a metal–metal bond unsupported by bridging ligands.

By circa 1955 binary carbonyl compounds of several metals had become generally available as laboratory reagents for the first time. The lability of their CO groups stimulated

Friends in München, 1978: Left to right, Gerda Beck, Liselotte Behrens, Joan Kaesz, who was visiting from Los Angeles with Herb Kaesz (not shown), Helmut Behrens, and Wolfgang Beck. Professor Behrens collaborated with Walter Hieber for almost 25 years at the Technische Hochschule in München and during this period was associated with many of the seminal early studies on metal carbonyls, including anionic carbonyl metallates. He was subsequently appointed to the chair of inorganic chemistry at Erlangen, and for many years his group was responsible for several significant discoveries, including that of the bridged-hydrido species $[M_2(\mu\text{-}H)(CO)_{10}]^-$ (M is Cr, Mo, or W). Today one accepts without question the presence of hydrido ligands bridging metal—metal bonds, but this was not so in 1957. Professor Wolfgang Beck, now at the University of München, was also a student of Walter Hieber and in early work demonstrated the value of IR spectroscopy as an analytical tool for the identification of carbonyl complexes. For many years Wolfgang Beck's group at the University of München, producing a stream of very important discoveries in the metal carbonyl area, have continued the tradition first set by Walter Hieber at the Technische Hochschule, now the Technischen Universität München.

numerous studies of reactions with various types of unsaturated organic molecules. Thus, with alkynes, iron carbonyls afforded a plethora of stable compounds. These species contained one, two, or three iron atoms, with organic fragments formed by linking of alkyne groups, a process often accompanied by incorporation of CO molecules within the ring systems.[16] The results illuminated the earlier work of Reppe and co-workers[17] on the

Lawrence F. Dahl (left) and Eugene R. Corey, of the University of Wisconsin, examine a molecular model of $[Rh_6(CO)_{16}]$ (Chem. Eng. News **1963**, *April 5). The structure and indeed the composition of this compound were unambiguously established by the single-crystal X-ray diffraction study. The molecule consists of six rhodium atoms located at the vertices of an octahedron. Each metal atom has a similar localized environment, being ligated by four adjacent rhodium atoms, two terminal carbonyl groups, and two of the four carbonyls that bridge alternate faces of the octahedron. The molecule contains 86 cluster valence electrons, but the significance of this number in metal cluster chemistry was not appreciated until the development of Wade's rules (J. Chem. Soc., Chem. Commun.* **1971**, *792). In other important work in this period, Dahl and Corey showed that the widely accepted dinuclear nonacarbonyls $[M_2(CO)_9]$ (M is Os or Ru) reported by the Hieber school had been incorrectly formulated and were in fact trimetal compounds $[M_3(CO)_{12}]$. The importance of these early X-ray diffraction studies cannot be overemphasized. Metal carbonyl complexes have played a pivotal role in the growth of transition metal organometallic chemistry. Moreover, the need for X-ray diffraction work in support of chemical syntheses became abundantly clear as more and more molecules with unusual structures were discovered.*

nickel salt catalyzed conversion of ethyne into cyclooctatetraene and on the formation of hydroquinones by treatment of alkynes with aqueous alkaline solutions of pentacarbonyliron.

By 1955 there was also renewed interest in the synthesis of metal complexes in which alkyl or aryl groups are σ-bonded to transition elements. A very timely review by Al Cotton[18] served to focus attention on this topic. Prior to 1955 there were few authenticated compounds of this class known, apart from methylplatinum compounds such as $[PtIMe_3]_4$.[19] Key discoveries made near the mid-1950s were the syntheses of the compounds $[TiPh_2(\eta^5\text{-}C_5H_5)_2]$,[20] $[WMe(CO)_3(\eta^5\text{-}C_5H_5)]$,[21,22] $[MnMe(CO)_5]$,[23] and $[FeMe(CO)_2(\eta^5\text{-}C_5H_5)]$.[22] The isolation of these and related complexes with methyl or phenyl groups led at that time to the belief that carbon–metal σ-bonds would be stable only if there were also present in these compounds other ligands possessing both donor and acceptor bonding properties, so that a synergistic effect between these two bonding modes could occur, enhancing the robustness of the molecule as a whole. As mentioned later in the book, with the preparation in more recent times of numerous isoleptic metal–alkyl and metal–aryl compounds, it is now appreciated that the need for the presence of π-bonding groups is not a priori essential and that the apparent instability of carbon–metal σ-bonds is associated more with kinetic than with thermodynamic factors. Thus these bonds are "stable" if present in molecules in which the metals are coordinatively saturated, and there are no low-energy pathways for decomposition.

Although Zeise's salt $[KCl\cdot PtCl_2\cdot C_2H_4\cdot H_2O]$ had been reported as early as 1825,[24] little was known about the chemistry of any transition element–alkene or transition element–alkyne complex before the appearance of a series of very influential papers by Joseph Chatt and his co-workers commencing in 1949.[25] The correct formulation of Zeise's salt as $K[PtCl_3(C_2H_4)]\cdot H_2O$ came to be recognized after an initially incorrect proposal[26a,27] that it was an alkylideneplatinum complex $K[Pt(=CHMe)Cl_3]\cdot H_2O$.[26b] Over the years the compound has been the subject of several X-ray and neutron diffraction studies, precise data being reported in 1975.[28] In 1953, Chatt and Duncanson[29] gave a definitive account of the properties of the anion $[PtCl_3(C_2H_4)]^-$ based on the concept of synergistic multiple bonding between ligated ethylene and the platinum(II) ion. This thesis is generally accepted to this day, but was based on an earlier seminal proposal of Michael Dewar[30] that filled d orbitals of the transition metals have the correct symmetry to overlap with the vacant antibonding orbitals of alkenes so that a second dative bond could be formed, opposite in direction to the σ donation from the filled bonding orbital of the alkene to a vacant metal orbital.

A Change of Fields

My entry into the field of organotransition metal chemistry was somewhat accidental. As mentioned earlier, as a graduate student and as a postdoctoral associate my training had been in the area of the boranes. When I joined Gene Rochow's group at Harvard I had intended to use this opportunity to gain experience in organosilicon chemistry, a field in which he was an authority, following his seminal work at the General Electric Company on the commercial synthesis of the methylchlorosilanes, precursors to the silicones. However, after I became an instructor in September 1954 I received considerable support and encouragement from Gene Rochow in setting up an independent research program. Initially, as mentioned, this program focused primarily on the interaction of various donor molecules with BH_3, BF_3, and BMe_3[31] and was in collaboration with my graduate students Bill Graham and Herb Kaesz, both now well-known chemists and working at the University of Alberta and at the University of California at Los Angeles, respectively.

Considerable intellectual stimulation was also provided by an association with Dietmar Seyferth, now at the Massachusetts Institute of Technology, who was a member of Rochow's group. At that time Professor Seyferth was developing the chemistry of the vinyl compounds of tin and other main-group elements.[32] While together at Harvard, Dietmar Seyferth and I wrote a paper[33] on the influence of 3d orbitals on the chemistry of silicon compounds. This article was for several years very widely

Judy Stone and Professor E. G. Rochow at a Christmas party in the Mallinckrodt Laboratory, circa 1957. Gene Rochow, always generous with equipment and facilities, was of great assistance in the early days of my academic career. Prior to his Harvard days Gene had worked at the General Electric Research Laboratories, where he had invented an economic process for making methylchlorosilanes on a commercial scale, key intermediates in the synthesis of silicone polymers.

cited, because in 1955 the involvement of d orbitals in bonds formed by second-row elements was not as generally accepted as it is today. Lou Allred, later to be a professor at Northwestern University, was at that time a member of Gene Rochow's group, and he also contributed many original ideas for our research and that of others.

While Dietmar Seyferth and I worked at one end of the first floor of the Mallinckrodt Laboratory, at the other end of the building Geoffrey Wilkinson, ably assisted by his students, was carrying out research that was subsequently to gain him the Nobel Prize. The two groups often had brown bag lunches sitting in my laboratory. Geoff's favorite "sandwich" after ferrocene consisted of sardines held in a bread roll, while I preferred

*Herbert Kaesz enjoys working with a high-vacuum system at Harvard,
circa 1957. Professor Kaesz (UCLA) was my second graduate student,
the first being Professor W. (William) A. G. Graham (University of
Alberta).*

bananas, having been deprived of this fruit in Britain for five
years in World War II. I well recall the agony, very discreetly
disguised, displayed by Bill Graham, who is fastidious about
such matters, when Geoff Wilkinson persistently spilt oil from
sardine cans on Bill's desk. However, I am sure Bill would have
adopted a more charitable attitude if he had been able to anticipate Geoff's Nobel Prize.

In 1956 Harvard's chemistry department became engaged
in one of its frequent exercises of dispensing with the services of
several nontenured teaching staff, and a vacancy for an assistant
professor in inorganic chemistry arose following Geoffrey

Bill Graham is put to work mowing my lawn in Bristol in 1988. No matter how well known an ex-graduate student may become, he or she must be re-employed whenever the opportunity arises.

Wilkinson's departure to Imperial College in January 1956. His leaving marked a lack of recognition at that time of the original-ity of the results that he and his co-workers had obtained and a general reluctance on the part of most tenured faculty to have any respect for synthetic inorganic chemistry. However, some inorganic chemistry had to be taught at Harvard, even though its renaissance was evidently not appreciated by several senior colleagues. I was therefore appointed to the assistant professor post vacated by Wilkinson, with duties that included the teach-ing of a graduate course in advanced inorganic chemistry. The lecture course attracted a number of excellent students and for balance naturally had to have both a transition metal and a main-group element content. In view of the paucity of suitable textbooks (the first edition of *Advanced Inorganic Chemistry* by Cotton and Wilkinson was not published until 1962, 5 years after I began to teach inorganic chemistry at the graduate level at Harvard) and the rapid advances being made, it was necessary to prepare most lectures directly from new information appearing in the primary literature or in reviews. The journal articles that I found the most exciting were those concerned with the synthesis of transition metal compounds having organic ligands. Moreover, after Wilkinson's departure from Harvard there was an obvious need for someone to take a research interest in what was evidently becoming a rapidly burgeoning field. Even so, I might not have begun work in this area had not R. B. (Bruce) King and T. (Tom) A. Manuel entered Harvard as graduate students in the fall of 1957. Early in 1958 they both joined my group and expressed a strong desire to carry out research on organotransition metal complexes. As I recall, Bruce King was the only graduate student I encountered at Harvard who had entered with the intention of researching in inorganic chemistry. All my other co-workers, with the exception of Bruce, had entered Harvard with the object of working with George Kistiakowsky, Frank Westheimer, Paul Bartlett, and other "stars" in the chemical firmament. The lack of graduate students with interests in inorganic chemistry was due in part to the fact that at that time the exciting new developments in the field were not being taught at the undergraduate level at the institutions from which the new graduate students came. However, my graduate course in inorganic chemistry changed the direction of several students away from organic and physical chemistry into organometallics, but hopefully in the longer term did not spoil their careers.

Cyclooctatetraene–Iron Complexes

There was no shortage of interesting problems in organotransition metal chemistry to address. For example, the ligating properties of cyclooctatetraene towards metal carbonyl fragments had not been tested. In 1958 an important paper by Hallam and Pauson[34] had appeared. It reported a reinvestigation of a product formulated as $[Fe(CO)_3(C_4H_6)]$, obtained much earlier by Reihlen et al.[35] from the reaction between butadiene and iron

Judy and I enjoy sea breezes on the Maine coast in 1958.

carbonyls. The new study showed that butadiene(tricarbonyl)-iron (1) should be regarded as a π or an η^4 complex,[36] with all four carbon atoms bonded to the iron.[34] Moreover, the new study focused attention on the superior bonding properties of conjugated dienes, as opposed to nonconjugated dienes, towards an iron tricarbonyl group.

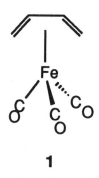

1

The Hallam and Pauson work[34] influenced us in the following way. The tub structure of cyclooctatetraene had been well established[37,38] by 1958, but it seemed to us that the hydrocarbon might react with iron carbonyls to bind two $Fe(CO)_3$ groups, one on either side of the ring, and adopt a new conformation so that two pairs of the four double bonds would form two "butadiene-like" systems, namely, **2**.[39] In practice the reaction of cyclooctatetraene with $[Fe(CO)_5]$ afforded a chromatographically separable mixture of three iron complexes formulated as $[Fe(CO)_3(C_8H_8)]$, $[Fe_2(CO)_6(C_8H_8)]$, and $[Fe_2(CO)_7(C_8H_8)]$.[40,41] The major product was $[Fe(CO)_3(C_8H_8)]$,[42] which ironically we would almost certainly have assigned correctly as the η^4 complex **3**, in view of the suggested structure for butadiene(tricarbonyl)-iron,[34] had we not had access to the Harvard chemistry department's 40-MHz ^1H NMR spectrometer. A primitive apparatus by the standards of today, the spectrometer was at that time "state-of-the-art" equipment. On recording the ^1H NMR spectrum of $[Fe(CO)_3(C_8H_8)]$ we observed a single sharp peak at room temperature, indicating an apparent equivalence of the eight protons. Variable-temperature studies were not possible on the equipment available, and the dynamic behavior of many molecules on the NMR time scale was not appreciated at

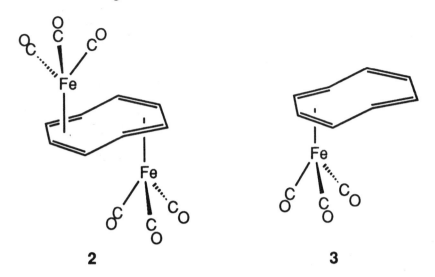

2   **3**

that time. To account for the unexpected observation of a single proton resonance, structure **4** was proposed, with a planar C_8H_8 ring attached to the iron—an idea, subsequently and amusingly with hindsight, supported by a theoretical treatment.[43] Recognizing the importance of establishing firmly the structure of [Fe(CO)$_3$(C$_8$H$_8$)] we suggested to W. N. Lipscomb that he carry out X-ray diffraction studies on both [Fe(CO)$_3$(C$_8$H$_8$)] and [Fe$_2$(CO)$_6$(C$_8$H$_8$)], it being anticipated that the latter would also contain a planar C_8H_8 ring with a Fe(CO)$_3$ group attached on either side. The X-ray results,[44] however, showed that the two compounds had structures **2** and **3**.

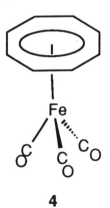

4

The observation of a single peak in the ^{1}H NMR spectrum of [Fe(CO)$_3$(η^4-C$_8$H$_8$)] was thus due to the now well-established phenomenon of fluxional behavior.[45] However, we might have arrived at the correct explanation for the observation of a single ^{1}H NMR signal had we fully appreciated the significance of an earlier paper by Piper and Wilkinson[22] in which they reported the synthesis of the iron complex 5. The ^{1}H NMR spectrum of

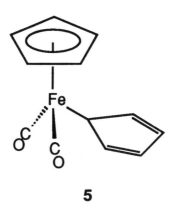

5

the latter at ambient temperatures showed two peaks instead of the more complex spectrum expected for the presence of a σ-bonded C$_5$H$_5$ group, with protons in different environments and an η^5-C$_5$H$_5$ ring. Piper and Wilkinson[22,46] proposed that rapid site exchange occurred, with the metal executing a series of 1,2-shifts around the monohapto-bonded cyclopentadienyl ring, a process leading to time averaging of the resonances. Had Tom Manuel and I made a similar suggestion to account for the ^{1}H NMR spectrum of 3 we would have had no means of experimentally verifying the idea in the fall of 1958, when the compound was first made. As mentioned, variable-temperature NMR facilities were not available, and even if they had been subsequent experiments with the iron species 3 showed that the equipment would have had to have operated at $\sim$−140 °C for any evidence of dynamic behavior to have been obtained. The fluxional behavior of 3 is associated with such a low activation energy that a limiting spectrum is not attained even at −155 °C. The problem was not conclusively solved until almost 10 years later when Al Cotton and his co-workers[47] studied the ruthenium complex [Ru(CO)$_3$(η^4-C$_8$H$_8$)], in the expectation, correct as it

Al Cotton photographed during a visit to Bristol in December 1991, when he was giving the inaugural lecture in a series to be given annually and founded to mark my time at the university. Al's researches on inorganic chemistry, for which he is known worldwide, have been reported in some 1200 papers. Without question a superb scientist with a brilliant intellect, his contributions have had a major influence on the development of organometallic chemistry, and several of his discoveries are mentioned at various places in this book.

turned out, that this species would have a higher activation energy for site exchange than the iron compound **3**. A limiting spectrum was reached for the ruthenium compound at ~-145 °C, and with the aid of computer-simulated spectra it was possible to show that the signal-averaging process was due to 1,2-shifts of the metal around the C_8 ring.

There are two interesting postscripts to this story. First, our observation of a single resonance in the ^{1}H NMR spectrum of **3**[39,40] generated considerable interest after Dickens and Lipscomb[44] had established the structure. Several groups prepared tricarbonyliron complexes of substituted cyclooctatetraenes in attempts to understand the fluxional behavior. Moreover, cyclooctatetraene(tricarbonyl)chromium, -molybdenum, and -tungsten complexes were prepared, and these species were also shown via ^{1}H NMR studies to undergo dynamic behavior in solution. An account of this work is given in the relevant sections of *Comprehensive Organometallic Chemistry*.[1] Second, although compound **3** does not contain a planar cyclooctatetraene ring, later we and others discovered several complexes in which the ligand adopts this geometry, for example, compounds **6**,[48] **7**,[49,50] and **8**.[51]

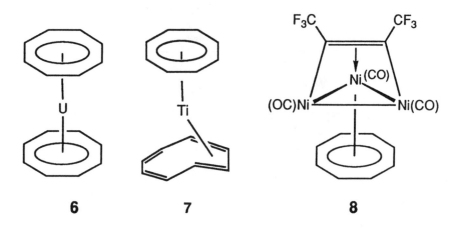

6   **7**   **8**

Extending the Scope of Metal π-Complex Chemistry

The range of alkene, diene, triene, and other π complexes available today must make it very difficult for the younger reader to appreciate how relatively few were known in 1958. In those early days one challenge arose as a consequence of the activities of Joe Chatt's very productive group at the I.C.I. Akers Research Laboratory. In a successful attempt to extend the scope of known alkene metal complexes, Chatt and Venanzi[52] prepared the rhodium compounds **9** and **10** and proposed that cycloocta-1,5-diene (cod) might well be a very suitable ligand to extend

9

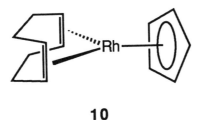

10

the range of metals known to form alkene complexes. Subsequent to the report of the rhodium complexes **9** and **10**, Roland Pettit[53] described the bicycloheptadieneiron and -molybdenum complexes **11** and **12**, the former being independently prepared by the Wilkinson group.[54] Pettit[53] had surmised that the two double bonds in bicycloheptadiene might be geometrically well disposed for metal bonding, while Wilkinson and co-workers[54] drew attention to the pseudo-conjugated-diene nature of the hydrocarbon, and related the ligand—metal bonding to that existing in the metal sandwich compounds. There is no doubt that Pauson's important paper[34] concerning the structure of $[Fe(CO)_3(\eta^4\text{-}C_4H_6)]$ (**1**) strongly influenced thinking at that time, with its emphasis on the conjugated double bonds in butadiene being the dominant factor for stability.

In order to establish beyond doubt that alkene complexes of molybdenum and tungsten could be obtained from dienes *where no conjugation between adjacent double bonds was possible*, Tom Manuel[55] prepared the cycloocta-1,5-diene-molybdenum and

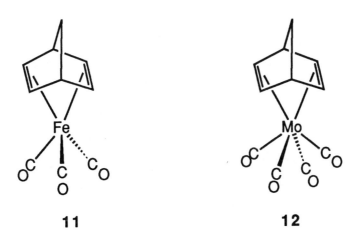

11   **12**

-tungsten complexes **13** by refluxing the hexacarbonyls of these metals with cod. In further work[55] designed to stress that two double bonds could "independently" ligate metals well to the left in the periodic table, dimethyldivinylsilane was used to obtain the compounds **14**. These were the first examples of metal–alkene complexes in which the double bonds are separated by a heteroatom. The complexes **13** were independently prepared by the groups of Fischer[56] and Wilkinson,[57] reflecting the competition at that time. Such has been the progress in organometallic chemistry; it will seem remarkable to younger readers that only some 35 years ago researchers were investigating whether any metals other than platinum would form complexes with alkenes.

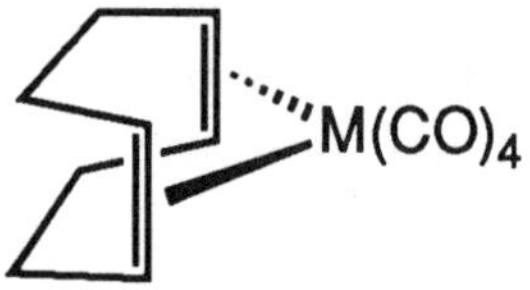

13 M = Mo or W 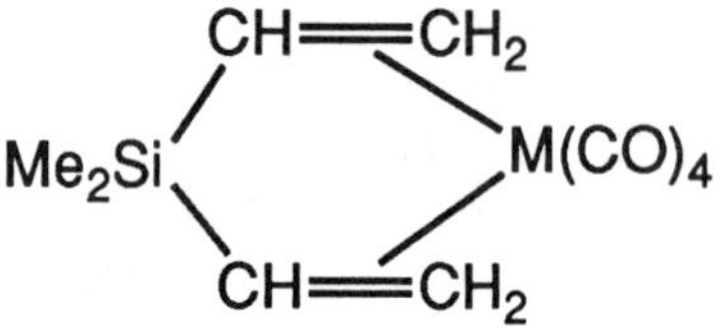   **14 M = Mo or W**

In attempting to prepare an iron tricarbonyl complex of a nonconjugated diene, we were somewhat less successful. From the reaction between $[Fe_3(CO)_{12}]$ and cod in benzene we obtained a product that we formulated as the desired compound **15**,[40,58] but with hindsight it is likely that we obtained a mixture of **15**[59] and **16**, since later workers[60,61] showed that both isomers exist, with, as expected, **16** being the more stable. Moreover, Arnet and Pettit,[62] in the same year as our work, reported that $[Fe(CO)_5]$ quantitatively converted cycloocta-1,5-diene to its 1,3-

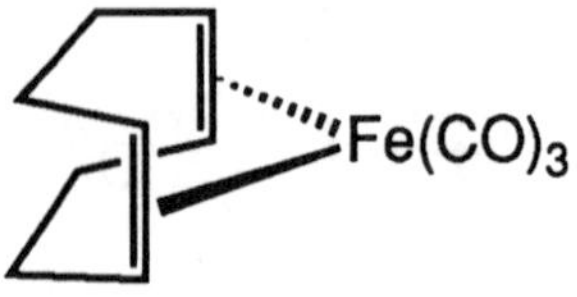

15    **16**

isomer. We also in 1961 reported[58] that treatment of penta-1,4-diene with $[Fe_3(CO)_{12}]$ resulted in isomerization of the diene, so that penta-1,3-diene(tricarbonyl)iron was formed. Much later, others showed that 1,3-diene(tricarbonyl)iron complexes are useful intermediates in organic syntheses,[63] with the metal carbonyl fragment being oxidatively removed at the end of a reaction sequence so as to liberate the desired organic compound. Indeed, Pettit[64] had pointed the way to this methodology with his elegant synthesis and studies on tricarbonyl(cyclobutadiene)-iron (**17**). The isolation of the stable compound **17** was a classic demonstration that unstable organic species could be captured by complexation with a transition metal. Sadly, Roland Pettit's early death removed one of the most brilliant practitioners of synthetic organometallic chemistry, as well as a fine personality.

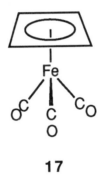

17

As a consequence of our own work on diene(tricarbonyl)-iron complexes another area of research opened up. Bruce King endeavored to use the divinyltin compound $Bu^n_2Sn(CH{=}CH_2)_2$ (Bu^n denotes n-butyl) as a chelate ligand towards an $Fe(CO)_3$ fragment, in the same manner as Manuel[55] had used Me_2Si-$(CH{=}CH_2)_2$ to obtain the molybdenum and tungsten complexes **14**. However, instead of producing a complex of structure **18**, the reaction between $[Fe(CO)_5]$ and $Bu^n_2Sn(CH{=}CH_2)_2$ yielded the metallacycle **19**, the vinyl groups being eliminated.[65,66] The ready formation of heteronuclear metal–metal bonds in this serendipitous manner provided the stimulus for research described in a later section, including the discovery of the novel spiro compound **20**.[67] Roald Hoffmann in his Nobel Prize lecture[68] pointed out that our pentanuclear metal compound **20** is an "inorganic" spiropentane, the tin atom replacing the central

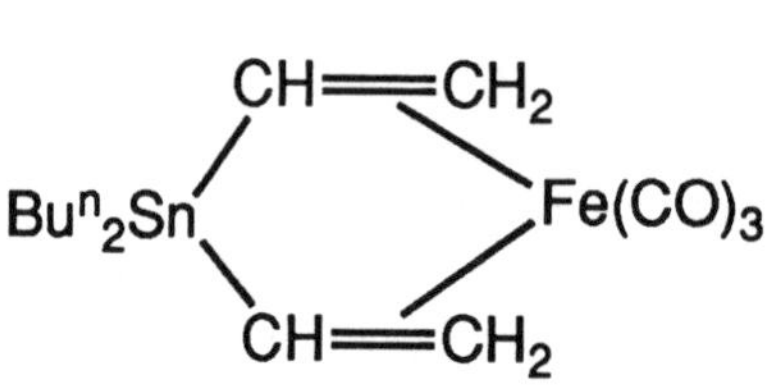

18 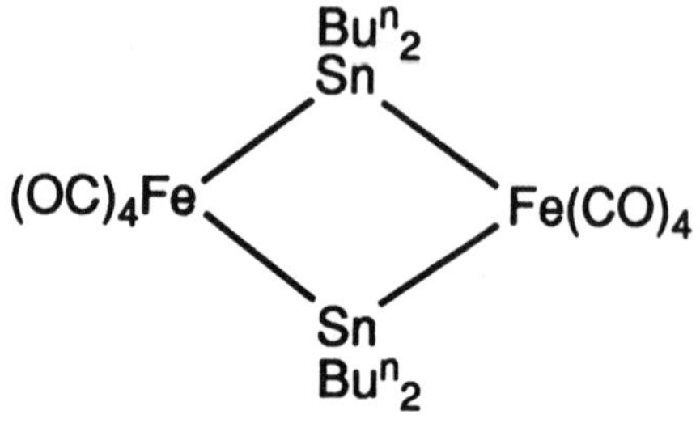   **19**

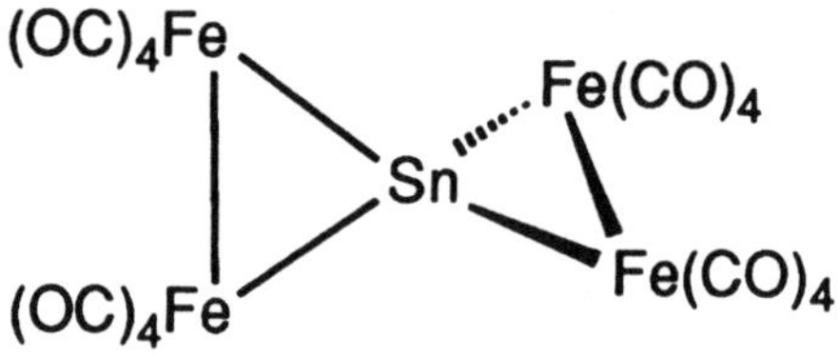

20

carbon atom in $C(CH_2)_4$, and $Fe(CO)_4$ fragments replacing electronically equivalent (isolobal, a term discussed later) CH_2 groups.

Another challenge in metal complex chemistry in the late 1950s, following characterization of species with η^5-C_5H_5 and η^6-C_6H_6 ligands, was the synthesis of compounds containing an η^7-C_7H_7 cycloheptatrienyl group. Since several metal cyclopentadienyl compounds had been prepared directly by heating metal carbonyls with cyclopentadiene, it seemed logical that $[M(CO)_x(\eta^7$-$C_7H_7)]_y$ species might be obtained similarly from cycloheptatriene and metal carbonyls. However, reactions of the triene generally afforded products containing C_7H_8 groups functioning as η^6-bound ligands, for example, the molybdenum complex **21**.[69] Success in obtaining an η^7-C_7H_7 metal complex was first achieved by Dauben and Honnen[70] in 1958 by hydride abstraction from **21** with $[CPh_3][BF_4]$, a reaction that afforded the salt **22**, formally containing the six-π-electron $C_7H_7^+$ cation.

In July 1959 Bruce King investigated the reaction between $[V(CO)_4(\eta^5$-$C_5H_5)]$ and cycloheptatriene in refluxing methylcy-

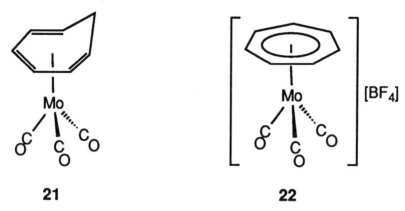

21   **22**

clohexane, in the expectation of displacing some or all of the carbonyl groups in the vanadium compound. The product of this
reaction was the purple crystalline sandwich compound **23**, the
first *neutral* η^7-C_7H_7 complex to be isolated.[71] There are some
amusing aspects to this discovery. Experiments were carried out
on July 16 and 17 of 1959, on the only weekend Bruce King
spent working in the laboratory at Harvard in the period September 1957 through June 1960, a success rate that should
encourage graduate students to work on weekends (Bruce, now
Regent's Professor at the University of Georgia, received the
American Chemical Society Award for Research in Inorganic
Chemistry in 1991). Also, the formulation we suggested for **23**,
the correct one as it turned out, was greeted with very considerable skepticism by certain senior colleagues on the staff at Harvard who preferred to believe that **23** was a bis(cyclopenta-

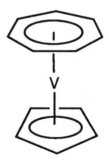

23

dienyl)vanadium compound in which one C_5 ring carried a vinyl substituent as a result of the C_7 ring collapsing to C_5.

At that time many of the new organometallic molecules did not have structures compatible with conventional thinking. Moreover, in 1959 chemists had not entered the era of X-ray crystallography as a practical means of cheap, rapid, and definitive analysis (*see* Al Cotton's and Jan Troup's classic paper on this subject[72]). However, I was able to persuade R. E. Rundle to carry out an X-ray diffraction study on our product **23**, and he and his co-worker G. Engebretson[73] confirmed our proposed structure *4 years* after the complex was first discovered. I have always been grateful to the late Professor Rundle for the very generous acknowledgment he expressed to us in his paper,[73] which served to alleviate the earlier skepticism. Currently in my laboratory we would expect to establish a structure of a potentially significant complex by X-ray diffraction within a few days of its first synthesis, *provided suitable crystals can be grown!*

In 1958 Fischer and Ofele[74] reported the chromium complex **24**, showing that thiophene could adopt an η^5 bonding mode to a metal. We studied[75] the reaction between thiophene and $[Fe(CO)_5]$ or $[Fe_3(CO)_{12}]$ in the expectation at that time of obtaining a complex $[Fe(CO)_2(\eta^5\text{-}C_4H_4S)]$.[76] Unexpectedly, the product was the diiron compound **25**, which had been reported a year previously as one of several complexes obtained from the reaction between acetylene and iron carbonyls.[77] The removal of sulfur from thiophene in its reaction with iron carbonyls is of interest in relation to the desulfurization of fossil feedstocks, an

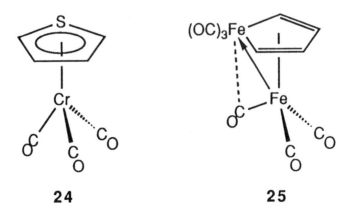

24   **25**

important industrial process catalyzed heterogeneously by metals.[78] Our work with thiophene was to my knowledge the first example of sulfur removal being accomplished by an organometallic reagent. As such it stimulated my group to study reactions between different types of organosulfur molecule and various carbonylmetal compounds. In this manner, cleavage of C–S and S–S bonds was observed, and numerous organothiometal complexes were isolated. Formulae **26–30**[79] give some impression of the scope of this study, which, together with the isolation earlier of complexes of type **31** by Hieber and co-workers,[80] provided the first examples of what subsequently has become a large class of metal compounds with μ-SR groups and cyclopentadienyl or carbonyl ligands. It remained for Bruce[81] to be the first to demonstrate that stereoisomers of such complexes could be separated in the pure state: the isomerism involves inversion at the sulfur atoms.

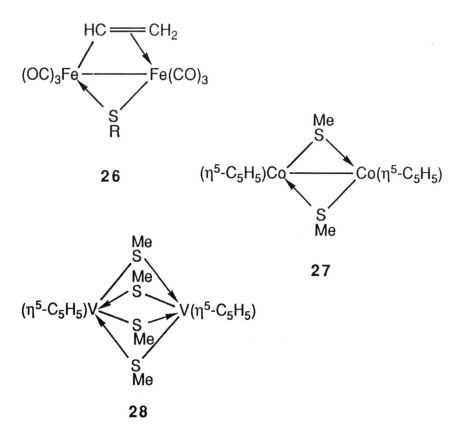

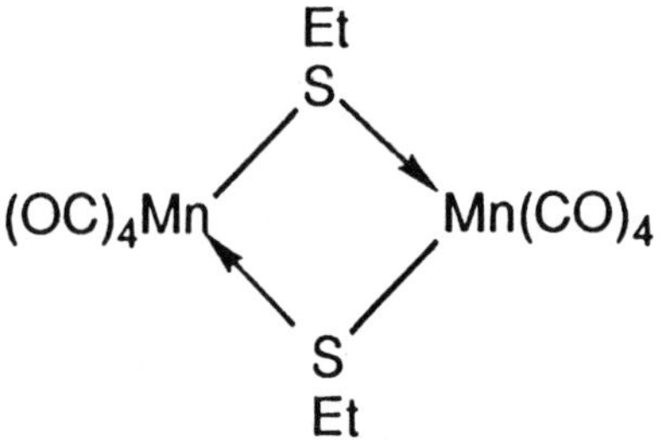

29

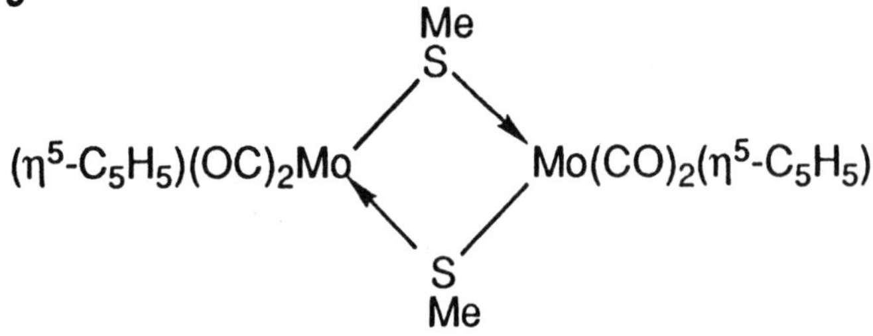

30

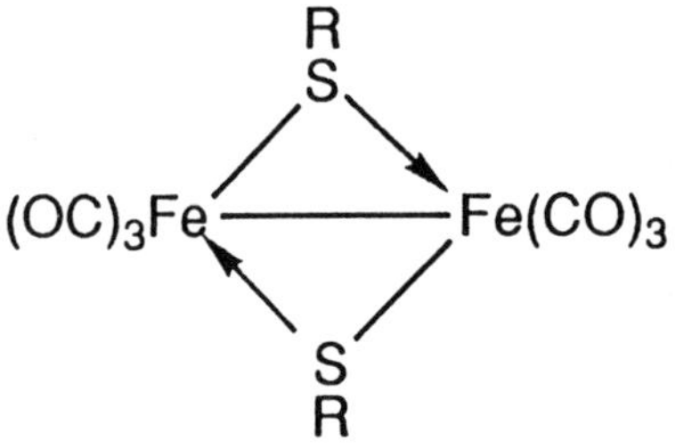

31 R = alkyl

The complexes **26** are worthy of special mention in the development of this area. They were prepared by heating vinyl sulfides RSCH=CH$_2$ with [Fe$_3$(CO)$_{12}$] in benzene[79a] and were, as far as I am aware, the first reported examples of compounds wherein a metal–metal bond is spanned by a vinyl group in a μ-σ,η^2-bonding mode. Only in recent times has this structural feature been shown to be common in organotransition metal chemistry, and that under some conditions the vinyl group rearranges to a μ-CMe fragment. In this early work was the beginnings of later research developed by many groups on the chemistry of small organic fragments at dimetal centers. Moreover, we used ^{1}H NMR spectroscopy to establish[79a] the presence of the vinyl group in the complexes **26**, and the observation of three distinct signals for the three protons with appropriate couplings (H–H trans > H–H cis > H–H gem) has been used subsequently by many others as a diagnostic tool for identifying μ-CH=CH$_2$ ligands.

Finally, it is interesting to recall that I insisted that Bruce King analyze the compounds **26** through reaction with trifluoroacetic acid. Ethylene and carbon monoxide were evolved quantitatively and were measured so as to establish the correct formulation of these species. Because I had carried out my postdoctoral research with Anton Burg, I was well equipped to train my students in the application of high-vacuum techniques to organotransition metal chemistry. In the absence of X-ray equipment to establish structures and mass spectrometry to establish molecular weights, our use of vacuum systems not only to manipulate volatile reagents, but also to analyze gases evolved in degradation experiments, enabled us to avoid the errors in stoichiometry and formulation of products somewhat prevalent in those days from some groups.

Allylmetal Complexes

Early in 1960 Heck and Breslow[82] at the Hercules Powder Company reported that they had synthesized allyl(tricarbonyl)cobalt (**32**) by treating allyl bromide with Na[Co(CO)$_4$]. This was an important result because it established clearly for the first time the trihapto- or η^3-bonding mode of an allyl group to a metal center. The first π-allylcobalt complex had in fact been prepared

32

in 1952 from butadiene and $[CoH(CO)_4]$, but the molecular structure of the reaction product was seemingly not recognized.[83] A somewhat similar situation existed with palladium. An allyl complex of this metal had been reported in 1959,[84a] but the true nature of η^3-allylpalladium complexes in general was not appreciated until somewhat later.[84b]

By 1960, as part of his Ph.D. program, Bruce King was exploring the reactivity of several anionic metal carbonyl complexes, and following a gift of $[Mn_2(CO)_{10}]$ from the Ethyl Corporation we were able to prepare $Na[Mn(CO)_5]$. Herbert Kaesz and Bruce King studied the reaction between allyl chloride and $Na[Mn(CO)_5]$ and thereby obtained the σ-bonded allyl complex $[Mn(CH_2CH=CH_2)(CO)_5]$. As expected, when the latter was heated in vacuo, carbon monoxide was released and the π-allyl manganese complex 33 was formed.[85] Today we would regard this reaction as trivial, but this was the first observation that the allyl ligand could adopt either an η^1 or an η^3 bonding mode with a transition element. The ability of organic fragments to display variable electron-donor properties towards a metal center by altering the number of carbon atoms bonded to the metal—for example, η^1 to η^3 (allyl) or η^3 to η^5 (cyclopentadienyl or indenyl)—is now known to be important in many organic syntheses involving the transition elements. However, this property was not well appreciated until more recent times. Our results on the allylmanganese complexes[85] were submitted in July 1960. In September of the same year workers at DuPont submitted a paper[86] describing several η^3-allyl(tricarbonyl)cobalt complexes, as well as the η^1-allyl(pentacarbonyl)- and η^3-

33

allyl(tetracarbonyl)manganese compounds. Thus the $\eta^1-\eta^3$ relationship for allyl ligands was arrived at independently by two groups. Competition between different laboratories has always been a feature of organometallic chemistry and has undoubtedly contributed to the vitality of this field. Very probably the need to obtain patent coverage delayed publication of the DuPont results on the allylmetal complexes, and in this respect workers in academia are at an advantage, since they do not have to obtain clearance from supervisors before submission of results for journal publication.

Fluorocarbon Complexes of the Transition Elements

Reference was made earlier to the situation prior to circa 1960, when relatively few transition element complexes were known with alkyl or aryl groups bonded to a metal. Thus while the number of π complexes of the transition elements was increasing in a seemingly exponential manner, knowledge of the σ-organometallic compounds remained limited. Several groups were to redress the balance. Chatt and Shaw,[87] at the I.C.I. Akers Laboratory, prepared many alkyl and aryl complexes of platinum as well as those of some of the other later transition elements by treating complex halides with organolithium or Grignard reagents, or by "oxidative addition" reactions (discussed later). The Pt^{II} and Pt^{IV} compounds $[PtMe_2(PEt_3)_2]$, $[PtI(Me)(PPh_3)_2]$, $[PtPh_2(AsEt_3)_2]$, and $[PtI_2(Me)_2(PEt_3)_2]$ are representative of many prepared at that time. The rationale behind the work was that the ligands PR_3 or AsR_3, with their π-acceptor properties, would stabilize the complexes by engaging in $d_\pi-d_\pi$ bonding with those filled d orbitals of the metal that are not involved in the formation of σ-bonds. It was argued that with all the d electrons occupied in synergistic bonding, the initial step of decomposition—promotion of an electron from a bonding to an antibonding orbital—would be made less favorable.

Somewhat earlier, Piper and Wilkinson[22,88] had prepared species such as $[CrMe(NO)_2(\eta^5-C_5H_5)]$, $[MEt(CO)_3(\eta^5-C_5H_5)]$ (M is Mo or W), and $[FePh(CO)_2(\eta^5-C_5H_5)]$. Another very important study in this period involved workers at the Ethyl Corporation. Treatment of $Na[Mn(CO)_5]$ in tetrahydrofuran with acyl halides afforded the acylmanganese complexes [Mn(COR)(CO)$_5$] (R is alkyl or aryl). These complexes decarbonylated on heating to yield the compounds [MnR(CO)$_5$],[23] species that have subsequently played a pivotal role in our understanding of the mechanism of alkyl migration processes.

Attracted to this area, we adopted a different approach to the synthesis of metal complexes with stable carbon–metal σ bonds. I recalled that during my graduate studies at Cambridge with Harry Eméleus, other members of his research group had discovered perfluoroalkylmercury compounds with seemingly remarkable properties[89,90] for organomercurials. Thus Hg(CF$_3$)$_2$ was a white solid, in contrast with liquid HgMe$_2$. Moreover, the perfluoroalkylmercurial compounds did not function as perfluoroalkylating reagents, whereas alkylmercurials, as is well known, transfer their alkyl groups readily to other metals. The properties of Hg(CF$_3$)$_2$, which set it apart from HgMe$_2$, are due to the electronegativity of the perfluoroalkyl groups; the electron-withdrawing property of a CF$_3$ group is similar to that of a chlorine atom. The perfluoroalkylmercurial compounds are thus in many respects more similar to mercury halides than to organomercurial compounds. I therefore reasoned that an analogous situation should occur with the transition elements, with fluorocarbon–metal species existing and having stabilities similar to well-established metal complex halides.

By 1960 my group at Harvard had grown in size, and several graduate students could be deployed to investigate whether or not fluorocarbon–metal complexes could be obtained. Employing the same method that had been used by the Ethyl Corporation group[23] to obtain the alkyl- and arylmanganese complexes [MnR(CO)$_5$], Kaesz and King[91] prepared perfluoroalkylmanganese and -rhenium pentacarbonyls:

$$R_FCOCl + Na[M(CO)_5] \xrightarrow{thf} [M(COR_F)(CO)_5] + NaCl$$
$$[M(COR_F)(CO)_5] \xrightarrow{heat} [MR_F(CO)_5] + CO$$

where M is Mn or Re and R$_F$ is C$_2$F$_5$ or C$_3$F$_7$.

Unknown to us at the time, the workers at the Ethyl Corporation had already reported the compound [Mn(CF$_3$)(CO)$_5$] at the First International Conference on Coordination Chemistry held in London in April 1959.[92] Abstracts of this meeting did not become available to us until after we had completed our studies. The Ethyl Corporation workers observed that $[Mn(CF_3)(CO)_5]$ did not react with CO to give $[Mn(COCF_3)(CO)_5]$, unlike $[MnMe(CO)_5]$, which readily affords $[Mn(COMe)(CO)_5]$. This result was in accord with our expectation of enhanced stability for a CF_3-Mn versus a CH_3-Mn bond, but the significance of the inertness of $[Mn(CF_3)(CO)_5]$ towards CO was evidently not recognized. In a paper submitted a month later than our own article,[91] W. R. McClellan[93] of DuPont reported the independent synthesis of several cobalt and manganese complexes, $[CoR_F(CO)_4]$ and $[MnR_F(CO)_5]$, via reactions between the carbonyl metal anions and perfluoroacyl halides. It was noted that cobalt compounds showed greatly enhanced thermal stability compared with their very labile $[CoR(CO)_4]$ analogs.

Within a few months my co-workers (T. (Tom) D. Coyle, H. D. Kaesz, R. B. King, D. W. McBride, P. (Peter) M. Maitlis, J. H. Morris, T. A. Manuel, J. R. Phillips, S. L. Stafford, and P. (Paul) M. Treichel) had developed several alternative routes to fluorocarbon–metal complexes.[94] Some of the preparative methods are shown in Figure 1. Important structural information was provided by IR and ^{19}F NMR studies carried out by Emily Pitcher (now Dudek).[95] Emily was much helped in her interpretation of IR spectra by attending a stimulating course on group theory given by Al Cotton, who by that time had joined the staff at MIT.[96] The ^{19}F NMR spectra of our compounds displayed interesting chemical-shift and spin-coupling relationships, and these data were interpreted with the assistance of A. D. (David) Buckingham, who was a lecturer at Oxford at that time, but was visiting Harvard for a few months. He was born in Sydney, my wife's home town, and this, coupled with his enthusiasm for all aspects of chemistry, brought us together. He and I developed a firm friendship that later (1965) influenced in part his becoming the first occupant of the chair of theoretical chemistry at Bristol, although he later moved on to Cambridge.

In those days recording ^{19}F NMR spectra was by no means routine and required resetting the magnetic field and other instrument adjustments before signals could be observed. Happily my students Tom Coyle and Emily Dudek became very adept with the early-model 40-MHz Varian NMR spectrometer

A photograph of some members of my group at Harvard taken on my 35th birthday, May 19, 1960. Left to right, back row: J. R. Phillips, F. E. Brinckman, T. A. Manuel, R. B. King, T. D. Coyle, P. M. Treichel, S. L. Stafford. Front row: L. D. Nichols, H. D. Kaesz, me, Emily Pitcher (later Dudek), and D. W. McBride. Absent from the photograph are W. A. G. Graham, who had left after acquiring his Ph.D., and postdoctorals Peter Maitlis and John Morris. Because Bill Graham's initial employment was at Arthur D. Little Inc., in Cambridge, Massachusetts, he continued to maintain contact and contribute to our research after he completed his Ph.D. degree.

available in the Harvard department. In other useful ^{19}F NMR work Tom identified labile mixed halides of boron, for example, BBrClF, in mixtures of BBr_3, BCl_3, and BF_3.[97] This was one of the earliest examples of redistribution studies of covalent halides with NMR spectroscopy as the analytical technique.

The fluorocarbon–metal compounds, as anticipated, revealed a greater robustness towards thermal or oxidative decomposition than their alkylmetal analogs where direct comparisons were possible. However, in the 30 or so years that have elapsed since their discovery, chemists have gained a much better understanding of the factors responsible for the stability

Varied Reactions Use Fluorocarbons Instead of Alkyl Compounds

1 C_3F_7COCl + $Mn(CO)_5Na$ $\xrightarrow{\text{Tetrahydrofuran}}$ $C_3F_7COMn(CO)_5$ + NaCl

   Heptafluorobutyryl chloride   Sodium manganese pentacarbonyl

   $C_3F_7COMn(CO)_5$ $\xrightarrow{\text{80° C.}}$ $C_3F_7Mn(CO)_5$ + CO

      Heptafluoropropylmanganese pentacarbonyl

2 $CF_2{=}CFCF_2Cl$ + $NaFe(CO)_2C_5H_5$ $\xrightarrow{\text{Tetrahydrofuran}}$ $CF_2CF{=}CFFe(CO)_2C_5H_5$

3 $[(C_6H_5)_2PCH_2CH_2P(C_6H_5)_2]Ni(CO)_2$ + C_3F_7I $\longrightarrow$

4 $Fe(CO_5)$ + $CF_2{=}CF_2$ $\longrightarrow$

      Iron pentacarbonyl   Tetrafluoroethylene

5 $HMn(CO)_5$ + $CF_2{=}CFCl$ $\longrightarrow$ $HCFClCF_2Mn(CO)_5$

Fluorocarbon Derivatives Form with Transition Metals That Don't Usually Yield Metal Alkyls

$C_3F_7Re(CO)_5$

Heptafluoropropylrhenium pentacarbonyl
white solid melting at about 27° C.

$(C_3F_7)_2Fe(CO)_4$

Bis(heptafluoropropyl)iron tetracarbonyl
pale yellow solid melting at 88° to 90° C.

Black solid decomposing at 145° C.

Yellow solid melting at 53° to 54° C.

Yellow solid melting at over 220° C.

Yellow liquid freezing below 0° C.

(trans Structure determined by nuclear magnetic resonance)

Figure 1. Discovery of fluorocarbon complexes of the transition elements, reported in an article in the October 2, 1961, issue of Chemical & Engineering News. *Copyright 1961 American Chemical Society.*

of metal—carbon bonds. Although thermodynamic factors can be important, as mentioned earlier, the avoidance of low-energy pathways for decomposition is essential. This necessity is well illustrated by the isolation during more recent times of homoleptic alkyls such as $[Ti(CH_2SiMe_3)_4]$, $[TaMe_5]$, $[Cr\{CH(SiMe_3)_2\}_3]$, $[WMe_6]$, and $[Re_2Me_8]^{2-}$, principally by the groups of Michael Lappert and Geoffrey Wilkinson.[98] With these species, decomposition via the so-called β-hydrogen elimination pathway or by other mechanisms is not possible. The stabilizing role of the π-acceptor ligands (CO, PR_3, η^5-C_5H_5, and so forth) is now recognized to be related to their strong occupancy of the coordination sites required for β-elimination or other processes to proceed.

Studies on fluorocarbon transition element compounds helped to focus attention on the properties of molecules with carbon—metal σ-bonds. Moreover, among some of the reactions we discovered and the products we obtained, one can find the origins of now well-established principles of organometallic synthesis. Examples are as follows:

1. Metallacycles are now recognized as important intermediates in many organic syntheses involving the transition elements. Our compounds **34** and **35** have been correctly described as the first reported examples of metallacyclopentanes.[99] The making and breaking of C—C bonds is of crucial importance in many homogeneously catalyzed reactions, and in order to understand the mechanisms of these processes it is valuable to isolate and establish the structures of the intermediates involved. Many of the fluorocarbon—metal compounds served as models for what came later.

34   **35**

2. The formation of the compounds **34** and **35** from C_2F_4 and [Fe(CO)$_5$] and from C_2F_4 and [Co(CO)$_2$(η^5-C$_5$H$_5$)], respectively, as well as the preparation of the complexes **36** and **37** by treating these same carbonylmetal complexes with C_2F_5I, led us to point out a close similarity between the reactivity patterns shown by [Fe(CO)$_5$] and [Co(CO)$_2$(η^5-C$_5$H$_5$)].[100] We also drew attention to similarities in the then-developing chemistry of the following pairs of compounds:

-   [Mn$_2$(CO)$_{10}$] and [Fe$_2$(CO)$_4$(η^5-C$_5$H$_5$)$_2$]
-   [Co$_2$(CO)$_8$] and [Ni$_2$(CO)$_2$(η^5-C$_5$H$_5$)$_2$]

We and others were just beginning at that time to recognize that certain metal–ligand fragments were electronically equivalent. Bruce King had appreciated that this property linked the three pairs of metal–ligand fragments: (1) Mn(CO)$_5$ and Fe(CO)$_2$(η^5-C$_5$H$_5$), (2) Fe(CO)$_4$ and Co(CO)(η^5-C$_5$H$_5$), and (3) Co(CO)$_3$ and Ni(η^5-C$_5$H$_5$). Hoffmann[68] subsequently placed these early nebulous ideas on a firm theoretical basis with his isolobal model, discussed later.

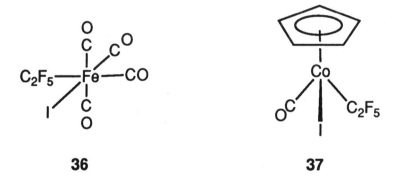

36   **37**

3. The syntheses of the compounds **36**, **37**, and [NiI(C$_3$F$_7$)-(Ph$_2$PCH$_2$CH$_2$PPh$_2$)] from reactions between perfluoroalkyl iodides and [Fe(CO)$_5$], [Co(CO)$_2$(η^5-C$_5$H$_5$)], and [Ni(CO)$_2$-(Ph$_2$PCH$_2$CH$_2$PPh$_2$)], respectively, provide early examples of pathways involving insertion of metals into carbon–iodine bonds, a process of fundamental importance in some catalytic processes. Chatt and Shaw[87] in their

studies on organoplatinum compounds had also observed
similar reactions in the preparation of $[PtI(Me)(PPh_3)_2]$
from $[Pt(PPh_3)_3]$ and MeI, and in the preparation of
$[PtI_2(Me)_2(PEt_3)_2]$ from $[PtI(Me)(PEt_3)_2]$ and MeI. Later,
these syntheses, carried out in Chatt's and in my laboratory, came to be classed as "oxidative addition" processes,
in which the metal is formally oxidized with an increase in
coordination number.[101] Only in the mid-1960s, following
the discovery of the reactions of the iridium(I) complexes
trans-$[IrX(CO)(PR_3)_2]$ (X is halide) with H_2, HX, RX, O_2, and
so forth did the concept of "oxidative addition" become
widely used to explain the ability of transition metals to
activate molecular hydrogen and many organic molecules
in both stoichiometric syntheses and catalytic processes.
Moreover, these later studies led Halpern[102] in a series of
seminal articles to identify relationships between particular
metal–ligand fragments and the free radicals, carbenes, and
carbanions of organic chemistry. It remained for Hoffmann[68] to account for these relationships on the basis of
the similarities in the frontier orbitals.

4. The reactions of hydrido(carbonyl)metal complexes with
    tetrafluoroethylene, giving species such as $[Mn(CF_2CF_2H)$-
    $(CO)_5]$ or $[Mo(CF_2CF_2H)(CO)_3(\eta^5\text{-}C_5H_5)]$,[94] were among
    the earliest well-characterized examples of the addition of
    transition metal hydrides to carbon–carbon double
    bonds,[103] a process now recognized as a key step in several
    homogeneously catalyzed processes.

It is amusing to recall that for these studies our supply of
C_2F_4 was obtained by the pyrolysis of Teflon polymer in a
Stock-type high-vacuum system. The Teflon was a gift from
DuPont, and I would like to acknowledge the friendship shown
to me by George Parshall and many other fine scientists at the
Central Research Department (CRD) of DuPont, demonstrated in
many ways during several visits over the years. The CRD
laboratory is always a stimulating place for discussions, and the
positive impact of the chemistry carried out there over the past
several decades is beyond measure.

In more recent times changes in policies have occurred, so
that regrettably more emphasis is being placed on research likely

to lead to commercially profitable products in the short or medium term. It is unfortunate for the future of American and British chemistry that chemical companies in both countries currently receive the attention of corporate raiders, and their actions, coupled with the rapacious demands of stockholders for increased dividend income, has forced the chemical industry to focus on short-term profits to the detriment of basic research. The present financial structure in Britain, with the presence of takeovers, emphasizes the role of accountants and has led to strategies that by their very nature do not include investments in research. Scientists need to play a bigger part in corporate decision-making.

Return to the United Kingdom

In September 1962, after 8 years at Harvard, I returned to Great Britain to join the staff of the chemistry department at Queen Mary College (QMC) of the University of London. Like many British people working overseas I had always thought in terms of returning to the United Kingdom at some stage in my career, but in my case this move was facilitated by the Harvard chemistry department's routine ejection of nontenured staff, especially if they had interests in synthetic inorganic chemistry. The enthusiasm of my Harvard research group, sustained in part by the institution of a daily tea break for discussions on the latest results, was greatly missed. I also missed the intellectual and social contacts with the numerous talented young chemists who were working in other groups in the department either as postdoctoral fellows or as other short-term visitors. It is not possible to say that I missed contacts with a friendly tenured academic staff. Apart from Gene Rochow, who with his wife Helen went to great lengths to put junior staff at ease, I had little social contact with senior colleagues during my time in Cambridge, quite in contrast with my experiences at USC.

My 8 years in the New England area had passed quickly. Most importantly I had acquired a wife and two sons. I had also developed close friendships with my Harvard graduate students and my postdoctoral assistants that will last all my life. Strong relationships are generated with those graduate students and postdoctoral fellows who work with one in the early stages of an academic career, probably because the age difference between supervisor and co-worker is not large. A photograph of most of my Harvard students appears in this book. However, three important people are missing from the photograph, namely my first postdoctoral fellows: Bodo Bartocha, Peter Maitlis, and John Morris. Bodo had come to Harvard from Marburg, and in our group discovered vinylmercury and vinylzinc compounds, which we used subsequently to prepare the first vinylboron compounds. Peter came from Michael Dewar's group, following a 1-year distraction as a postdoctoral assistant in organic chemistry at Cornell. I was glad to be able to introduce him to the then rapidly developing area of organotransition element chemistry, a field in which he has since remained. Peter became a professor of chemistry, first at McMaster and subsequently at Sheffield, and was elected to the Royal Society in 1984. John Morris came from Norman Greenwood's group at Leeds, and after working with me joined the staff at the University of Strathclyde in Glasgow, where he has published much good work on boron compounds. Bodo Bartocha, after periods working in various laboratories of the U.S. Navy, joined the National Science Foundation and served for many years as the Director of International Programs. He is currently a Professor of Arid Lands at the University of Arizona in Tucson.

At the time of my departure from the United States in the fall of 1962, Bob West and I became co-editors of the series *Advances in Organometallic Chemistry*. This collaboration arose through our having been approached independently by different people at Academic Press to act as the editor. I had met Bob when I first arrived at Harvard, when he was on the point of leaving Gene Rochow's group. Fortunately, we had remained in contact after his departure, and so in a conversation during one of our meetings we discovered that we had both been asked to become editors of the series. Since our research interests were complementary and we could reasonably cover all of organometallic chemistry it seemed like a perfect plan to act as co-editors, and this collaboration has worked out splendidly, with 35 volumes appearing so far. I wonder how the publisher would have resolved the situation had we both individually accepted our respective invitations to edit the series.

Dinner parties for graduate students were popular occasions for members of my Harvard group and their wives, not least because of Judy Stone's cheesecake and other dishes. This page, top: Judy Stone, Elisabeth Bartocha, and Meg Brinckman. This page, bottom, left to right: Tom Manuel, Herb Kaesz, Joan Kaesz, and Bill Graham. Facing page, top: Paul Treichel and Emily Pitcher Dudek; subsequently Professor Treichel (Wisconsin) and Professor Dudek (Brandeis).

I arrived at Queen Mary College about 1 year after Michael Dewar had departed for Chicago, and so unfortunately I did not have the benefit of interacting with him at my new institution. I had first met Michael Dewar when I gave a seminar at Chicago in 1962, and after my talk he kindly drove me to

Dinner at Sheffield to celebrate the election in 1984 of Peter Maitlis to the Royal Society. Left to right: Peter Maitlis, Marion Maitlis, me, and Diane McCleverty.

The rapid development of organometallic chemistry in the 1960s led to the initiation of a series of International Conferences held at Cincinnati (1963), Madison (1965), München (1967), Bristol (1969), and Moscow (1971). These meetings have continued on a biannual basis for many years and have recently been supplemented by a series of international conferences on the application of organometallic compounds in organic synthesis. The photo shows, from left to right, me, E. O. Fischer, Robert West, and Ray Dessy at the Madison Conference in 1965, which was organized by Bob West.

Midway airport, carrying on a stimulating conversation all the way. Fortunately I arrived safely at the airport.

    Both Judy and I had to adjust to some strange attitudes and rules in Britain in the fall of 1962. For example, I shipped my car from the States to England with us as accompanied baggage. The car was an early "compact" model, but even so it dwarfed its British counterparts in size, and I suppose my driving it was regarded by some as a "status symbol" long before this term was in common usage. I had learned to drive in Los Angeles some 10 years earlier and had even driven from Los Angeles to Boston before the age of interstate highways. Even so, as a returning U.K. citizen the law stated I must pass a British driving test. Moreover, according to the law, I was supposedly not allowed to drive the car without a British license after sunset on my first day of arrival. Since those of visitor status were allowed to drive without a U.K. license, I pretended to be a visiting American for several months, which was not difficult with my mid-Atlantic accent. Eventually I took the driving test, which proved an interesting experience. In those days seat belts were unheard of for cars. With the examiner I had driven less than 50 yards down a narrow street when a woman carrying a considerable amount of shopping suddenly appeared from behind a parked car on the road directly in front of us. I trod vigorously on the brakes, stopping two or three feet from her, but the examiner, who was sitting on my right, cracked his head on the windshield, fortunately not hurting himself seriously. After he had regained his composure he quickly got out of the car and issued me with the necessary document for a driving license. I never paid off the lady who caused me to brake so suddenly.

In October 1962 the facilities for research at QMC were primitive, as indicated in one of the accompanying photographs. However, plans for a new building were well advanced, and it was constructed in the mid-1960s shortly after I had departed. The winter of 1962–1963 was exceedingly cold. The Thames froze, which is extremely rare, and London was experiencing the last of the great breath-choking fogs that many Americans, as a result of watching old British movies on late-night TV, still believe represent a permanent state of the English winter climate. One afternoon I was working in the college library when I became aware that it was becoming difficult to see the bookshelves at the end of the large reading room. Having established to my amazement that this was due to fog seeping into the building, I had to consider how it might be possible to travel home to Loughton, the suburb of London in which we then lived. In such circumstances the underground railway system was the obvious means of escape from the East End of London. With visibility on the street down to about 6 feet, and with the nearest station some half a mile from Queen Mary College, this presented a problem. However, by several of us holding hands to form a chain, and groping our way, we eventually reached the Mile End underground station, colliding with others and with buildings. Fortunately, as a result of the clean air laws, such air pollution has disappeared from the British cities,

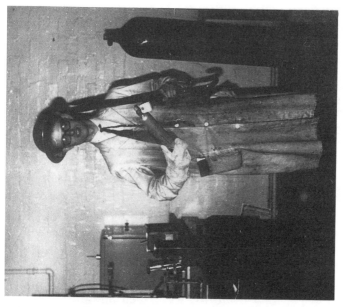

The transfer from Harvard to Queen Mary College (University of London) resulted in considerable disruption of my research. As indicated by the photograph on the left, substantial experimental work was required of me in bench reconstruction! In surmounting these difficulties, Ph.D. students provided assistance, as shown in the photograph on the right, where John Wilford adopts the spirit of the London Blitz of 1940—1941 to deal with inadequate and hazardous laboratory facilities.

but it took me a few days to recover from the experience. My wife frequently recalls that I left her to recover our car from the QMC car park the next day, no mean accomplishment since the fog had only partially lifted. Being a resourceful Australian she was able to cope.

Paul Treichel enthusiastically suggesting an idea at a mandatory tea break, circa 1960 (see text).

Paul Treichel, one of the last of my Ph.D. students at Harvard, made the move from Cambridge to London with me and thus, with the aid of his experience and skills, the momentum of research, although much reduced for a period, was not entirely lost. Moreover, I was immediately joined in London by three excellent British graduate students: Peter Jolly,* David Rosevear from Joe Chatt's laboratory, and John Wilford, a graduate of Queen Mary College.

Our research at Queen Mary College was directed at preparing transition metal complexes in which perfluoroaromatic groups were bonded to the metals. This was a logical extension of the perfluoroalkylmetal chemistry described; moreover, by selecting this area I knew my research output would suffer least from the disruption caused by the move from Harvard. Success was quickly achieved with the isolation of such species as $[M(C_6F_5)_2(\eta^5-C_5H_5)_2]$ (M is Ti or Zr), $[Re(C_6F_5)(CO)_5]$, $[Ni(C_6F_5)_2(PPh_3)_2]$, $[Pt(C_6F_5)_2(PEt_3)_2]$, and many similar compounds.[104] Other groups quickly took up the study of fluorocarbon–metal compounds so that by 1968 more than 500 were known,[105] and many more have been characterized in the past 25 years. An extensive chemistry has now grown around the C_6F_5 derivatives of gold and platinum, an area elegantly developed by the Spanish group of Professor Rafael Usón,[106] several of whose students have worked with me.

Inorganic Chemistry at Bristol

I remained at Queen Mary College for only 10 months. The early 1960s was a period of very rapid expansion in the British university system, due to a massive injection of additional government funding. The number of universities was almost doubled, and the older institutions increased in size as student intake rose. Consequently, in August 1963 I moved to Bristol University to become the first occupant of a newly created chair of inorganic chemistry. Previously there had been only chairs of organic chemistry and physical chemistry at Bristol, although within a year of my arrival a fourth chair, in theoretical chemistry, was established. As was mentioned earlier, the first occupant of the latter was my friend David Buckingham.

Apart from the advantages to be gained professionally by moving to Bristol, a large city located in one of the most

---

* After completing his Ph.D. work with me and a postdoctoral appointment with Roland Pettit, Peter Jolly joined the staff of the Max-Planck-Institut für Kohlenforschung at Mülheim, where he has become well known for his work on organonickel chemistry, not least by his scholarly survey of the extensive organic chemistry of nickel in reference 1.

My strong links with Spanish laboratories led to my making several visits to that country. Generous hospitality was always provided, as in this photograph of wine sampling with Rafael Usón in about 1984.

pleasant areas of the United Kingdom, there was a very considerable gain in the quality of life for my wife, my sons, and for me. Unlike the situation in London, which has become increasingly more difficult as the years have gone by, it was possible to find good housing within a 5- to 10-minute drive from the laboratory, thereby allowing me ready access to my office at the weekends, a time for work regarded as unsocial by my neighbors who were employed in the many commercial organizations in the city. I have always appreciated living close to my office, thereby eliminating time wasted on commuting. Moreover, ready access to the laboratory facilitated better contact with co-workers. Making oneself available to graduate students and postdoctoral assistants, even outside normal working hours, I deem to be of crucial importance for their well-being. Indeed I was able to design my office suite at Bristol in such a way that co-workers could gain access to me without having to pass by the secretaries.

When I arrived in Bristol there were only four or five members of staff with interests in mainstream inorganic chemistry. Among these, however, were three key colleagues: E. W. Abel, N. S. Hush, and P. Woodward, whom I was glad to support. Some years later, Noel Hush (deservedly elected to the Royal Society in 1988) left to become foundation professor of theoretical chemistry at the University of Sydney. Peter Woodward was a great asset to the department, setting up and supervising our X-ray laboratory, and for many years, until his early retirement in 1983, giving sound advice on many aspects of teaching and administration. Although chronologically out of order, it is worth mentioning that Peter's departure, as well as that of many others in the British university system, was linked to the steady attrition in the financial resources available, a situation that developed in the 1980s. Ironically, it was often the case that it was one's more talented colleagues who availed themselves of the opportunity to leave when funding was reduced, rather than those with tenure who had little to show in research. Since those who departed were not replaced, the overall quality of the academic staff remaining was not necessarily raised by the various financial inducements held out to encourage people to retire prematurely. This result of course was contrary to what the bureaucrats who organized these early retirement schemes probably expected, since it would seem to them that the loss of academic staff, brought about by financial constraints, would give the authorities in the various universities a mechanism to eliminate staff who were unproductive. Considerations based on the fact that universities had contractual obligations to staff through tenure were seemingly forgotten, at least in the initial stages. In more recent times tenure has been removed for all new academic appointments and for those staff in post if promoted to a higher grade.

The third member of the trio at Bristol when I arrived was Edward Abel, who is well known among organometallic chemists. He had an active research group working in a very old laboratory when my co-workers and I arrived from London, and our groups shared this laboratory during the construction of a very well equipped new building, which was completed during 1964. The advent of the new building and a large increase in undergraduate student numbers during the next decade allowed the appointment of several additional teaching staff in inorganic chemistry. Among early appointments were those of Peter Goggin from Oxford, Michael Green from the University of Manchester Institute of Science and Technology, and Peter Timms, lured back to England from Berkeley.

A year or so prior to my arrival in Bristol, the chemists were very fortunate in acquiring an excellent person, Alan Honey, as the laboratory administrator. He was able to help the professors in numerous ways, in dealing with the central administration, technical staff, safety matters, and a host of other problems. Alan, who was a chemist by training, had gained valuable previous experience working for the Department of Scientific and Industrial Research, an organization that many years later evolved into the U.K. Science and Engineering Research Council. Shortly after I met Alan I told him "I will do the chemistry and you sort out the central administration", and he amply fulfilled his part of the bargain. For 27 years I could not have wished for a better or more congenial colleague. Alan took early retirement when I moved to Texas, a step I interpreted in part as a compliment to our relationship, although more truthfully it was associated with his desire to lead a more relaxed life-style as the academic scene became more bleak.

The expansion of the British university system that took place in the 1960s brought about changes in the way in which many departments of chemistry were managed. Prior to this period, with most departments having but one or perhaps two chairs of chemistry, and with limited equipment, there was generally a pyramid structure, with the other teaching staff (lecturers and readers) doing research in collaboration with a professor. The creation of multichair departments, with the commensurate addition of many appointments at the lecturer level, hastened the creation of a situation whereby all faculty members were expected to manage their own research. I strongly encouraged this trend among both old and new inorganic chemistry staff at Bristol by providing all possible support in the form of graduate students, equipment, and laboratory space so as to allow research programs independent of my own to flourish. My efforts in this direction were gratifyingly marked over the following years by the success of the Bristol inorganic chemistry staff in gaining appointments to chairs.*

I also devoted considerable effort in setting up a strong laboratory infrastructure for research. For example, Robin Goodfellow, who came to Bristol initially as a postdoctoral fellow from Venanzi's group at Oxford, was appointed to organize our NMR service and to train students in this technique. In setting up service facilities for NMR and mass spectrometry to be made widely available to graduate students and postdoctoral assistants, I was enthusiastically supported by Mark Whiting, who arrived to occupy the chair of organic chemistry, following the retirement of Wilson Baker. Mark had been at Harvard at the time the sandwich structure of ferrocene had become recognized, and he had been a coauthor with Bob Woodward and Myron Rosenblum on a communication about this discovery. Mark Whiting appreciated, as did I, that we had entered an era where those engaged in synthesis had to be supported by ready access to spectroscopic techniques, and the time had passed when such methods were the sole purview of the physical chemist. Although well appreciated today, the need for inorganic and organic chemists to have ready access to instrumentation had only begun to be recognized in the late 1950s and early 1960s, and Mark Whiting and I had to adopt a strong stance with some senior colleagues to ensure that our views were accepted.

It also was necessary to establish a weekly research colloquium with guest speakers from other British universities and

---

* Edward Abel remained at Bristol as a major contributor to the life of the department until January 1972, when he moved to Exeter to become that University's first professor of inorganic chemistry. Six persons from the inorganic chemistry department at Bristol moved to chairs at other universities during the period I was head of department. In addition to Abel and Hush, the others were M. I. Bruce (to Adelaide), J. L. Spencer (to Salford), M. Green (to King's, London, and then Bath), and Judith A. K. Howard (to Durham). Moreover, the university wisely appointed S. A. R. Knox to a chair at Bristol, following my departure in 1990. As far as I am aware, no other inorganic chemistry group in the United Kingdom generated as many professors as did Bristol during the period I was there. I found it particularly frustrating that after about 1980 it became impossible for financial reasons to replace those talented chemists who had left. The difficulties would have been easier to bear if senior administrators had appreciated that it was the better departments that were being undermined by the policies developing from about the mid-1970s.

from overseas, and to make sure that the graduate students and final-year undergraduate students attended the talks. Such actions may seem strange to mention today, but at that time there was a laissez-faire attitude, and I was determined to create an enthusiastic and competitive atmosphere in building up the department so that chemistry could be taught and practiced at the highest level.

Having become established at Bristol in 1963–1964, the small group I had brought from Queen Mary College was soon augmented by several other talented graduate students, from Bristol and elsewhere. These included in the early years Colin Cundy, Neil Mayne, Dick Osborn, Tony Rest, and Alison Taunton-Rigby. Later the research momentum was further accelerated by the arrival of Jonathan Ashley-Smith (now Scientific Curator at the Victoria and Albert Museum), Jane Browning (later Berry), John Clemens, and David Empsall, together with postdoctoral assistants Dave Cook and Peter Maples, the latter from Fred Basolo's group. All played important roles in developing the chemistry of fluorocarbon ligands. By 1965 tetrafluoroethylene was being used by us and by others as a model substrate molecule to probe the nucleophilicity of low-valent species typified by $[IrX(CO)(PPh_3)_2]$ and $[Pt(PPh_3)_4]$. This work formed part of a general developing interest in the ability of d^8 or d^{10} metal compounds to activate small molecules. George Parshall and colleagues at DuPont[107] had found that Vaska's complex $[IrCl(CO)(PPh_3)_2]$ formed an adduct $[IrCl(CO)(PPh_3)_2(C_2F_4)]$ on treatment with C_2F_4. They proposed that this product should be regarded as an Ir(III) species formally containing the $[C_2F_4]^{2-}$ dianion. The related rhodium complex $[Rh(acac)(C_2H_4)(C_2F_4)]$ (acac is acetylacetonato) was also prepared. In a later important study involving collaboration between workers at DuPont and John D. Roberts,[108] the complex $[Rh(C_2H_4)(C_2F_4)(\eta^5-C_5H_5)]$ was shown by NMR spectroscopy to undergo dynamic behavior in solution involving rotation of the C_2H_4 group about an axis through the rhodium and the midpoint of the C=C bond. In contrast the C_2F_4 ligand remained rigid, reflecting its tighter bonding to the metal. Interestingly, John Roberts was a coauthor of the work described in reference 108. His autobiography, *The Right Place at the Right Time*, has been published in this series, and in his book he has given an entertaining account of discussions held in 1954 at the Central

Relaxing in Bill Graham's garden after the Boomer lectures, University of Alberta, 1965.

Research Department of DuPont about ^{19}F NMR spectroscopy, when the technique was in its infancy. Evidently, as a visitor to DuPont at the later time (circa 1969)[108] he was assisting in interpreting the spectra associated with the dynamic behavior of the rhodium complex.

We focused initially on the reactivity of the Malatesta[109] compound [Pt(PPh$_3$)$_4$]. This compound on treatment with a variety of fluoroalkenes afforded the very stable complexes [Pt(fluoroalkene)(PPh$_3$)$_2$].[110]   In   solution   the   fluoroalkene ligands in these complexes do not rotate about an axis through the metal atom and the midpoint of the C=C bond, as do hydrocarbon   alkenes   in   the   compounds   [Pt(alkene)(PR$_3$)$_2$]. Moreover, ^{19}F NMR measurements carried out by Tony Rest revealed J_{FFgem} values for the CF$_2$ groups in the fluoroal-keneplatinum compounds similar to those observed in fluorocy-clopropanes. These coupling constants were much higher ($\sim$190 Hz) than expected ($\sim$60 Hz) for =CF$_2$ groups associated with sp^2-hybridized carbon atoms, a result suggesting that the complexes are best formulated as metallacyclopropanes with σ-bonds between platinum and the ligand, as depicted by **38a–38d**.

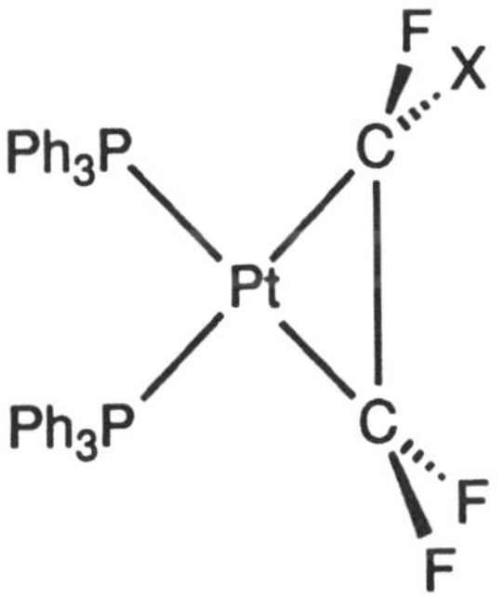

	X
38a	F
38b	CF$_3$
38c	Cl
38d	Br

According to the Dewar–Chatt–Duncanson model, discussed earlier,[29,30] this representation corresponds to a high degree of retrodative bonding between the filled d orbitals of platinum and the π^* orbitals of the fluoroalkene. This feature is probably responsible for the ready rearrangement of compounds **38c** and **38d**, either thermally or in polar solvents, to give σ-vinyl complexes, a process accelerated by silver ions:

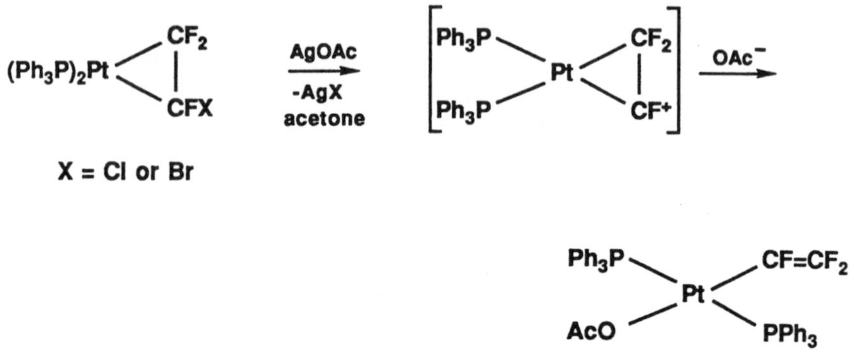

X = Cl or Br

Certain other Lewis acids also promote rearrangement to a vinylmetal complex. Thus **38b** on treatment with $SnCl_4$ affords $[PtCl\{C(Cl)=CF(CF_3)\}(PPh_3)_2]$, and $[Pt\{CF_2CF(CF_3)\}(Ph_2PCH_2-CH_2PPh_2)]$ with the tin reagent gives $[PtCl\{C(F)=CF(CF_3)\}-(Ph_2PCH_2CH_2PPh_2)]$.

Further developments included the discovery that not only fluoroalkenes but many different types of unsaturated organofluorine compound would react with the d^{10} $[Pt(PR_3)_4]$ complexes. Organofluoropalladium compounds were also isolated, and the availability of labile nickel complexes, such as $[Ni(cod)_2]$ (cod is cycloocta-1,5-diene),[111] enabled similar chemistry with this metal to be explored. Moreover, with the discovery of a convenient synthesis of $[Pt(cod)_2]$, discussed later, new reactivity patterns for Pt^0 were observed. Some idea of the scope of the work, covering the period circa 1965–1976, can be gained by reference to structural formulae **39–52**.[112] Certain aspects of this work merit mention.

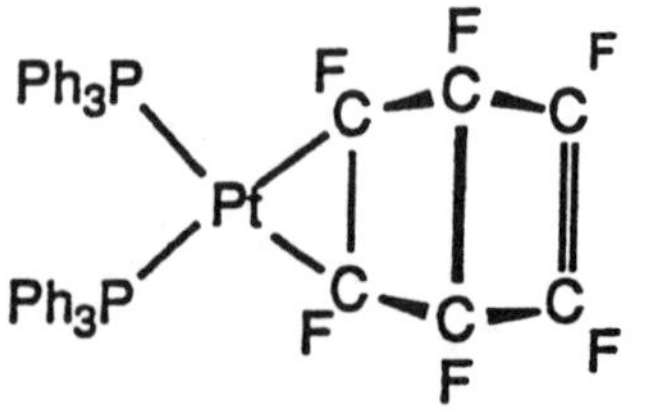

39

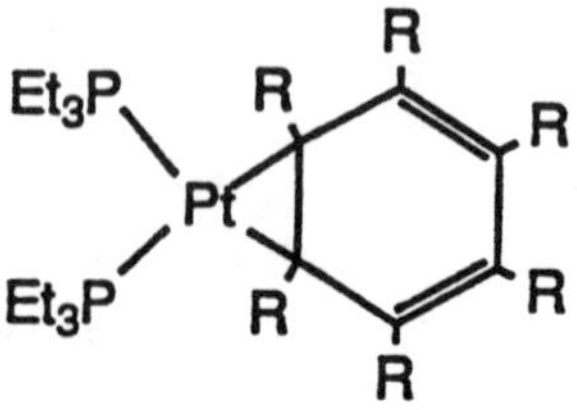

40 R = CF$_3$

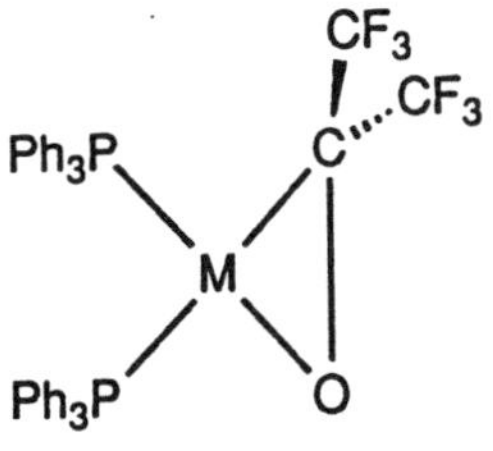

	M
41a	Ni
41b	Pd
41c	Pt

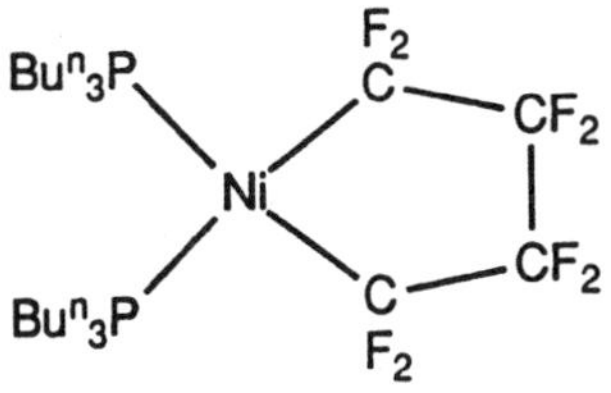

42

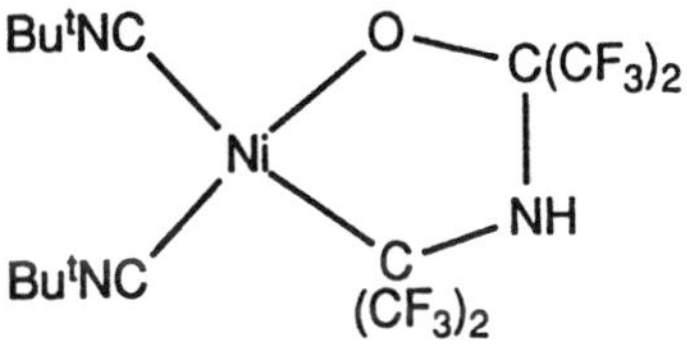

43

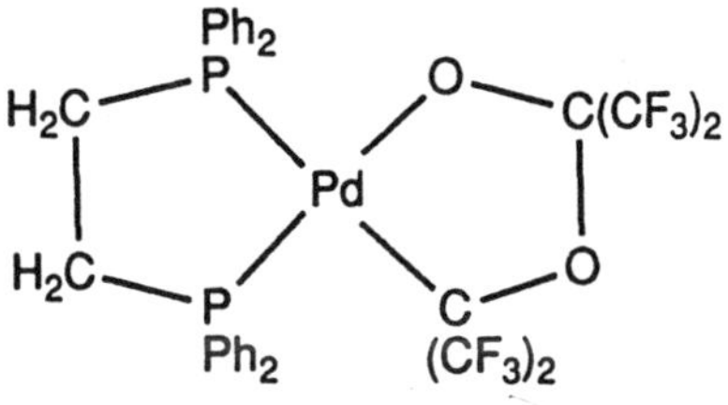

44

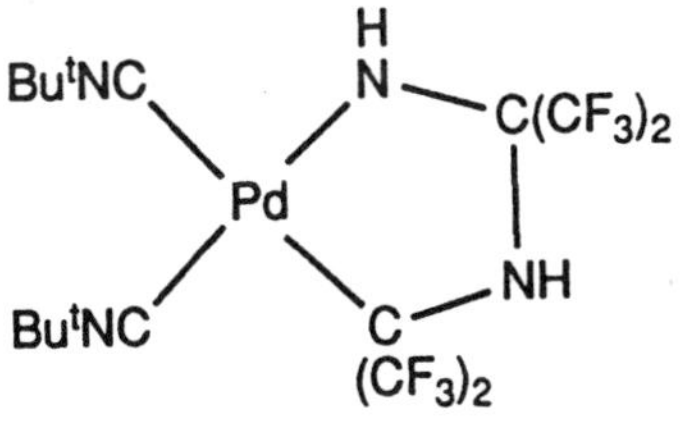

45

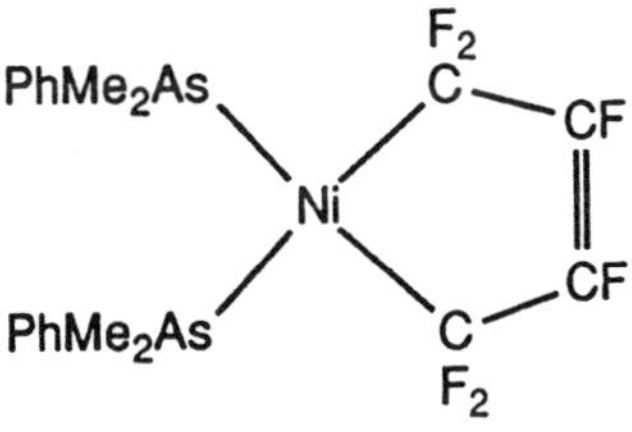

46

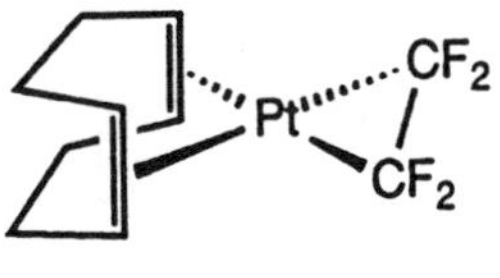

47

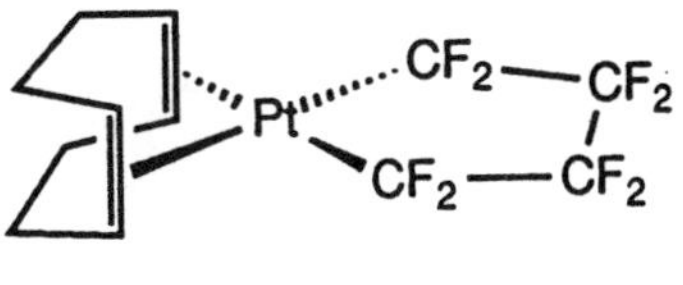

48

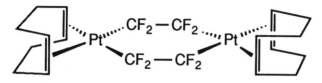

49

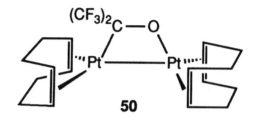

50

51 R = CF₃

52 R = CF₃

1. Compounds of considerable structural interest were characterized, for example, the Dewar benzeneplatinum compound **39** and the η^2-hexakis(trifluoromethyl)benzeneplatinum complex **40**. The latter was the archetype of mononuclear transition element complexes having η^2-ligating benzene rings.

2. The products obtained can differ widely, depending on the precursor used. Thus C_2F_4 reacts with [Pt{methyl(vinyl)-ketone}(cod)], [PtPr$^i_2$(cod)] (Pr^i is isopropyl), and [Pt(cod)$_2$] to give compounds **47**, **48**, and **49**, respectively, having very different molecular structures.

3. Many complexes containing five-membered rings were isolated. In some instances these products were obtained directly from reactions between zero-valent Ni, Pd, or Pt compounds and fluorocarbon substrates. For example, complex **42** was prepared by treating [Ni(cod)$_2$] with two equivalents of PBu$^n_3$ and C_2F_4, and compound **45** was synthesized from [Pd(CNBu^t)$_2$] (Bu^t is *tert*-butyl) and (CF$_3$)$_2$-C=NH. It was also shown that the five-membered metalla-ring complexes could be constructed in a stepwise manner via the intermediacy of three-membered ring species, for example, the metallacycle **43** from [Ni{C(CF$_3$)$_2$-O}(CNBu^t)$_2$] and (CF$_3$)$_2$C=NH or compound **44** from [Pd-{C(CF$_3$)$_2$O}(Ph$_2$PCH$_2$CH$_2$PPh$_2$)] and (CF$_3$)$_2$C=O. The formation of three- or five-membered ring compounds with the various fluorocarbon substrates could be controlled in many reactions by changing the metals or their attached ligands. Thus CNBu^t was superior to CNPh for activating nickel, and PMePh$_2$ was better than PPh$_3$ for activating palladium. In this manner it was demonstrated that stronger σ-donor ligands increased the nucleophilicity of the metal centers, promoting insertion of a second molecule of substrate. It was also shown that steric factors and vacant coordination sites played an important role in the various C–C, C–O, and C–N bond-forming processes observed.

4. Evidence was accumulated that the mechanisms of ring

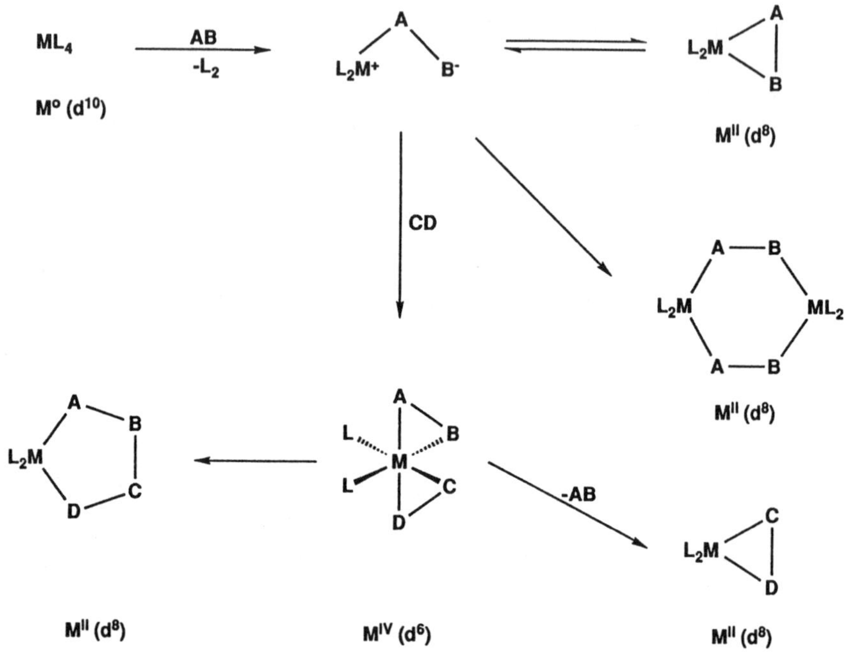

Scheme I. *Schematic representation of the synthesis of three-, five-, and six-membered metalla-ring complexes of Ni, Pd, or Pt, possibly via an initially formed dipolar intermediate.* **AB** *and* **CD** *are unsaturated fluorocarbon substrate molecules. In many reactions* **AB** *and* **CD** *are the same molecular species (e.g.,* $CF_2=CF_2$*).*

formation involved dipolar intermediates, shown here for the synthesis of **38a** (*see also* Scheme I):

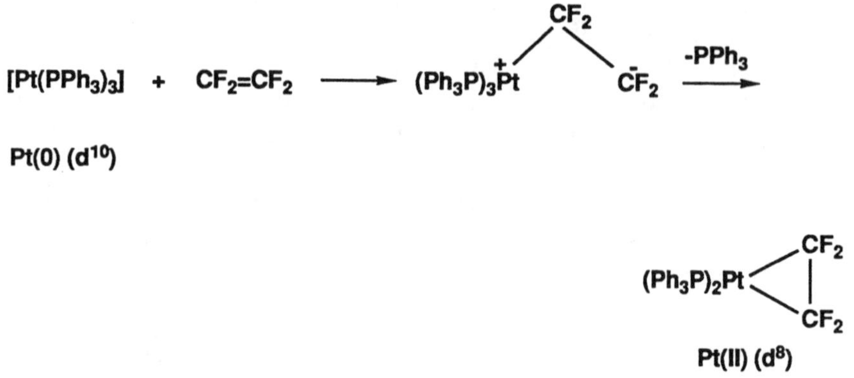

Moreover, it is interesting to compare this reaction with

$$[Re(CO)_5]^- \; + \; CF_2=CF_2 \; \longrightarrow \; (OC)_5Re \overset{CF_2}{\diagdown} \overset{}{\underset{\bar{C}F_2}{\diagup}} \xrightarrow{\;-F^-\;}$$

Re(-I) (d^8)

$$[Re(CF=CF_2)(CO)_5]$$

Re(I) (d^6)

The fluorocarbon substrates thus proved useful in establishing reactivity patterns and models for oxidative addition processes for d^{10} complexes. Although ^{19}F NMR spectroscopy was very useful in establishing the structures of the products, X-ray diffraction studies by Bruce Penfold and his co-workers (Christchurch, New Zealand), for example, on the complexes **40**, **41a**, and **43**, were of major importance for our research. I had met Bruce earlier at Harvard, and as a result a valued friendship was established.

Ruthenium Carbonyl, Silicon–Ruthenium Bonds, and Pentalene

By the time I arrived in Bristol in 1963 it was very apparent that X-ray diffraction had become an indispensable tool for organometallic chemists, and I was well aware of the need to set up a strong X-ray crystallographic group. In this I found an enthusiastic ally in my colleague Peter Woodward, mentioned earlier, who at my instigation shortly after my arrival at Bristol spent a period working in Professor L. F. Dahl's laboratory in Wisconsin. Larry Dahl has also been mentioned previously in this book, in connection with the establishment of the structures of [Rh$_6$(CO)$_{16}$] and [Mn$_2$(CO)$_{10}$], and over many years his detailed structural analyses have played a pivotal role in advancing organometallic chemistry.

The year 1963 was fortunate for me and for organometallic chemistry at Bristol in that both Judith A. K. Duckworth (later Howard) and Selby A. R. Knox arrived at the university as

Left to right: Paolo Chini, Gunther Wilke, Larry Dahl, and Peter Pauson at the Ettal Conference on Metal Carbonyls (1974). It would be difficult to assemble a group of four chemists who have made more significant contributions to the development of modern organometallic chemistry. Paolo Chini pioneered high nuclearity metal carbonyl cluster chemistry until his premature death in 1980. Gunther Wilke made the seminal discovery of "naked nickel" alkene complexes with the isolation of such species as cyclododecatrienenickel and developed their use as catalysts. Larry Dahl successfully elucidated the structures of $[Fe_3(CO)_{12}]$, $[CFe_5(CO)_{15}]$, and $[Rh_6(CO)_{16}]$, and many other significant molecules at a time when X-ray structural studies were not routine. Peter Pauson's name will always be associated with the discovery of ferrocene. The photograph is yet another taken at the landmark meeting held in Ettal, Bavaria, in 1974 to commemorate Walter Hieber's 80th birthday

undergraduate students. Their interest in organometallic chemistry came about through undergraduate research projects in crystallography and synthetic chemistry, respectively. Moreover, in 1965 Michael Bruce, an Oxford graduate, returned to England from Australia to work with me as a graduate student. Judith Duckworth, Selby Knox, and Michael Bruce were to play important roles in the chemistry described in this section, as was John Cotton, a postdoctoral assistant from Melbourne who is now

E. O. Fischer and me at the Walter Hieber Symposium (Ettal, Bavaria) in July 1974. It is possible that Professor Fischer is explaining the 18-electron rule, but if so his pupil seems more interested in the photo opportunity.

Ron Mason (left) and Mike Lappert, both well known in the field of organometallic chemistry, listen to Judy Stone making a point at the Ettal Symposium, July 1974. I am standing in the background.

Werner Fellmann (Wacker Chemie), Michael Lappert (Sussex University), Herbert Kaesz (UCLA), and Sergei Gubin (A. N. Nesmeyanov Institute of Organoelement Compounds, Moscow) at Ettal, Bavaria, 1974. Dr. Fellmann was one of several postdoctoral students from München to join Herb Kaesz's group at UCLA. Mike Lappert is known to all organometallic chemists for his many contributions to the field. Sergei Gubin was the organizing secretary for the 5th International Conference on Organometallic Chemistry held in Moscow in 1971, under the chairmanship of Academician A. N. Nesmeyanov.

chairman of the department of chemistry at the University of Queensland in Brisbane.

John had a good sense of humor, a trait common with those from the antipodes, and several years later he confessed to a good practical joke at my expense. Rules concerning the disposal of chemical waste were relaxed in that period so one day, having discovered that a worker on an adjacent bench had drained a foul-smelling oil from a vacuum pump, he distributed the oil under the previously mentioned American car I had brought back to the United Kingdom from Boston. It was running nicely between the wheels when I left the laboratory to go home at my usual late hour, and apparently the sight of my head under the hood examining the engine for the source of the oil was the cause of much hilarity among members of my group. Social and informal meetings of the group were frequent. Many parties were held at our home, but since we did not at that time possess a dishwasher, clearing up afterwards was a chore. One party coincided with a visit by a distinguished professorial colleague from Germany who assisted with the washing up but entreated us not to inform his wife when next we met her. I have respected his desire in this respect. Duties were also assigned to graduate students of that era, one of whom, Brian Goodall, wrote to me many years later, on the occasion of my 60th birthday, to remind me that on one occasion he had been assigned to run my dog around the block as a forfeit for some game. Brian, when subsequently working in industry, was responsible for the discovery of an important commercial process for producing polypropylene.

I had remained intrigued by the iron–tin compounds such as **19** prepared by Bruce King at Harvard. In 1965 John Cotton and Selby Knox, the latter through his undergraduate research project with me, were investigating reactions of organotin halides with $[Fe(CO)_4]^{2-}$ or neat $[Fe(CO)_5]$. Among several new tin–iron cluster compounds discovered was the novel species **20**[67] that, as mentioned earlier, later attracted Roald Hoffmann's interest.[68] The structure of compound **20** was determined by Judith Howard as part of her undergraduate research project, supervised by Peter Woodward. Later Judith was to do her doctoral work in Dorothy Hodgkin's group at Oxford. For-

Homework after a research group party at our house, 1966.

Brian Goodall takes time off from research to show Judy and me some Dutch scenery during a visit I made to the Netherlands to give a lecture at the Koninklijke/Shell-Laboratorium in Amsterdam.

tunately, she subsequently returned to Bristol, initially as a post-doctoral assistant but subsequently as a valued member of the academic staff, to play for more than 20 years a pivotal role in our X-ray laboratory. Shortly after my departure from Bristol she moved to a chair of structural chemistry at Durham, and, as far as I am aware, Judith is one of only two women professors of chemistry in the United Kingdom at the time of writing. While this small number is regrettable, I can at least take some credit for helping to improve the situation by 50%.

In an independent line of research, also begun in 1965, Michael Bruce discovered a convenient synthesis of ruthenium carbonyl (53). Michael's Ph.D. project was directed towards the synthesis of $[Ru_2(CO)_4(\eta^5\text{-}C_5H_5)_2]$, the chemistry of which was undeveloped at that time. We planned to prepare this compound by treating ruthenium carbonyl chloride with NaC_5H_5 or TlC_5H_5, but the formulation of the carbonyl chloride was obscure at the time. As described later, we initially characterized

Judith Howard, mentioned several times in this book, was associated with me at Bristol for many years in several capacities, from postdoctoral assistant to senior academic colleague.

53

the chloro-bridged diruthenium compound $[Ru_2Cl_2(\mu\text{-}Cl)_2(CO)_6]$ and its tetrahydrofuran adduct $[RuCl_2(CO)_3(thf)]$. As part of this study a simple synthesis of $[Ru_3(CO)_{12}]$ emerged. Prior to Bruce's work ruthenium carbonyl was not readily accessible. Mond et al.[113] had observed orange crystals, presumably $[Ru_3(CO)_{12}]$, as a consequence of treating ruthenium metal with CO under extreme conditions (~400 atm at 300 °C), but it was not until 1961, a year before I left Harvard, that the correct formula for the carbonyl was established by an X-ray diffraction analysis; this accomplishment represented yet another significant contribution from Larry Dahl to this field.[114] The pivotal importance of the compound as a reagent, and its significance in metal cluster chemistry in general, is indicated by its having been the subject of two additional X-ray diffraction studies, undertaken in order to obtain precise structural parameters.[115,116]

Michael's original synthesis[117] involved the low-pressure carbonylation of methanol solutions of ruthenium trichloride at ~60–70 °C in the presence of zinc as a halogen acceptor. Yields were as high as 80% if the solutions were recycled. In the absence of zinc, the dimeric ruthenium carbonyl halide $[Ru_2Cl_4(CO)_6]$ was formed. Within a few years, modifications of this synthesis were introduced by various workers, and several alternative methods of preparation were reported.[118] Moreover, so rapidly did the chemistry of $[Ru_3(CO)_{12}]$ develop that the compound became available from commercial suppliers. I remember suggesting to Mike Strem that his company market the carbonyl, and this was subsequently done. It would be presumptuous to claim that the early work at Bristol[119] was solely responsible for the subsequent research activity, since by the mid-1960s $[Ru_3(CO)_{12}]$ was a molecule, like its osmium analog, whose time had come! The chemistry of $[Fe_3(CO)_{12}]$, a compound that had been readily accessible for some years,[120] was well developed,[121] with its reactivity mainly associated with fragmentation of the iron triangle to afford products with $Fe(CO)_4$ groups. In contrast, it had become generally recognized that metal–metal bonds were stronger with the heavier transition elements. Consequently, it seemed likely that $[Ru_3(CO)_{12}]$ would be a source of polynuclear ruthenium compounds, as was shown in many later studies by Mike Bruce and others.[122] Alternatively, if the carbonyl were a source of $Ru(CO)_4$ groups, these fragments should form strong bonds either with main-group atoms or with transition elements.

With a good supply of $[Ru_3(CO)_{12}]$ available, several lines of investigation were initiated in our laboratory. The anion $[Ru(CO)_4]^{2-}$ was prepared and protonated to give $[RuH_2(CO)_4]$, and the reactions of this hydride were explored.[123] In another project, mixtures of $[Ru_3(CO)_{12}]$ and $[Fe(CO)_5]$ in scrambling reactions afforded the chromatographically separable carbonyl compounds $[FeRu_2(CO)_{12}]$, $[Fe_2Ru(CO)_{12}]$, and $[FeRu_3(\mu\text{-}H)_2$-$(CO)_{13}]$.[124] The latter species resulted from the presence of water in commercial $[Fe(CO)_5]$! This reaction was not the first and will not be the last where the use of contaminated reagents results in the unexpected formation of an interesting product. More importantly, this methodology for synthesis involving scrambling reactions between binary metal carbonyls to afford "mixed-metal" cluster compounds was widely adopted in later years by others to obtain many new metal clusters.[125]

It was because of the aforementioned results with the iron–tin compounds, leading to products such as **20**, however, that we were prompted to investigate reactions of $[Ru(CO)_4]^{2-}$ and $[Ru_3(CO)_{12}]$ with various organoderivatives of Si, Ge, and Sn. This area of study was taken up by Selby Knox and others as part of their graduate work. With $[Ru_3(CO)_{12}]$, depending on the reaction conditions, the hydrides Me_3EH (E is Si, Ge, or Sn), Me_5Si_2H, and $Me_4Si_2H_2$ afforded the mononuclear ruthenium species $[Ru(EMe_3)_2(CO)_4]$, as well as a wide range of di- and triruthenium compounds **54–57** containing organosilicon, organogermanium, and organotin groups.[126,127] Of these, the

	E
54a	Si
54b	Ge
54c	Sn

	E
55a	Si
55b	Ge
55c	Sn

56

	E
57a	Si
57b	Ge

mono- and diruthenium species were the most interesting. The complexes **54a–54c** are readily reduced to the mono anions $[Ru(EMe_3)(CO)_4]^-$. Like the anion $[Mn(CO)_5]^-$, to which it is electronically related, the ruthenium species was used to prepare compounds with metal–metal bonds, for example, $[Me_3Si(OC)_4RuMn(CO)_5]$ and $[Me_3Si(OC)_4RuAu(PPh_3)]$.

In 1972 Selby Knox returned to Bristol as a staff member after a period as a postdoctoral fellow at UCLA with Herb Kaesz, and at that time I was able to increase the strength of the inorganic chemistry area even further by the appointment of Neil Connelly to a lectureship. Selby and I engaged in a short but fruitful collaboration while he rapidly built up an independent group. With our co-workers we studied the reactions of cyclic polyolefins (for example, cyclohepta- and cyclooctatrienes, cyclooctatetraene and substituted cyclooctatetraenes, and cyclododeca-1,5,9-triene) with the compounds $[Ru(EMe_3)_2(CO)_4]$ and **54**, and numerous interesting products, for example, **58–62**, were thereby isolated. There are two important features of these syntheses: (1) transannular ring closure of a cyclic olefin as seen in the tetrahydropentalenyl complex **58**, and (2) migration of $SiMe_3$ or $GeMe_3$ groups from ruthenium to the hydrocarbon ligand as in **59–61**. It was, however, a reaction between $[Ru(GeMe_3)_2(CO)_4]$ and cyclooctatetraene that proved to be the most significant because it afforded a diruthenium complex **63b** containing the elusive hydrocarbon pentalene **64**.[128]

Since free pentalene can only be detected spectroscopically at very low temperatures, complex **63b** is yet another illustration of a situation in which a transient organic molecule can be stabilized as a ligand in a transition metal complex. There are

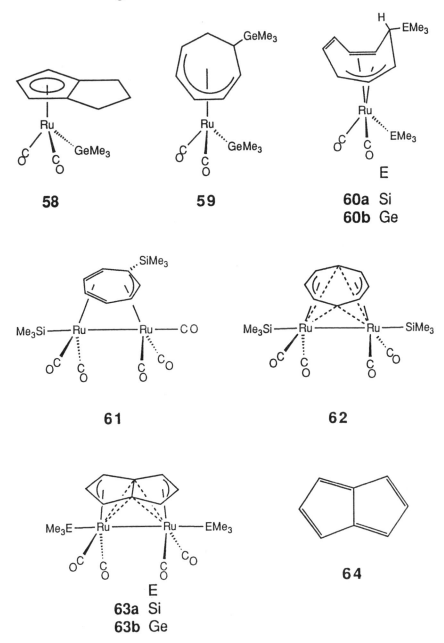

58   **59**   **60a** Si
                                60b Ge

61

62

63a Si
63b Ge

64

numerous examples known of the capture of normally unstable organic species by coordination to iron carbonyl fragments[129] that follow in the footsteps of Pettit's[64] seminal work leading to tricarbonyl(cyclobutadiene)iron (**17**).

The pentalene molecule in compounds **63a** and **63b** is formed by a dehydrogenative transannular ring closure of cyclooctatetraene, a process presumably facilitated by elimination of GeHMe$_3$. Subsequent, more detailed studies revealed that the mononuclear ruthenium compounds **60a** and **60b** are intermediates in the reactions of the reagents [Ru(EMe$_3$)$_2$(CO)$_4$] with cyclooctatetraene, which yield the pentalene complexes **63a** and **63b**.[130] There is tentative evidence that the conversion proceeds via the mechanism shown in Scheme II. The complexes **60a** and **60b** are fluxional; two enantiomorphs interconvert in solution. The oscillation in bonding shown in the following structure generates a time-averaged mirror plane of molecular symmetry. Two protons lie in the mirror plane, hence in the limiting high-temperature spectrum there are five ring-proton resonances, and at −60 °C there are eight.

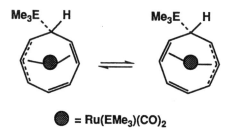

$\bullet$ = Ru(EMe$_3$)(CO)$_2$

Formation of the complexes **63a** and **63b** raised the intriguing possibility that [Ru$_3$(CO)$_{12}$] might also afford a pentalene complex by dehydrogenation of cyclooctatetraene. Reinvestigation of the reaction between [Ru$_3$(CO)$_{12}$] and C$_8$H$_8$ by Victor Riera, the first of several excellent postdoctoral colleagues from Spain and now a professor at the Departamento de Química Organometálica, Universidad de Oviedo, did indeed lead to the isolation of a triruthenium complex [Ru$_3$(CO)$_8$(η^8-C$_8$H$_6$)], with the pentalene ligand lying along the edge of the metal triangle, attached to two ruthenium atoms (Figure 2). Since this cluster compound was formed in yields of only 1 to 3%, it was not surprising that its existence had been entirely missed in the earlier work that had led to the characterization of the three major products of the reaction: [Ru(CO)$_3$(C$_8$H$_8$)], [Ru$_2$(CO)$_6$(C$_8$H$_8$)], and [Ru$_2$(CO)$_5$(C$_8$H$_8$)].[47,131] Considerable experimental skills

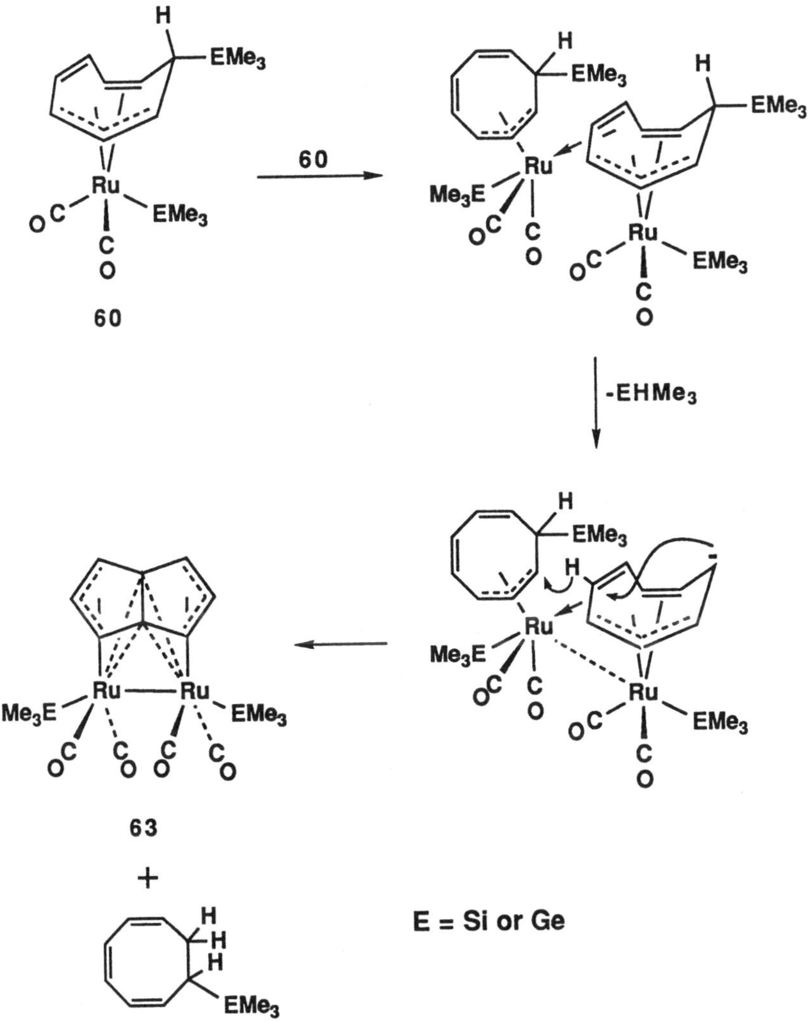

Scheme II. *Possible pathway to the pentalenediruthenium complexes* **63** *from the mononuclear ruthenium precursors* **60**. *(Reproduced with permission from reference 130a. Copyright 1978.)*

were displayed by Victor in the isolation of $[Ru_3(CO)_8(\eta^8\text{-}C_8H_6)]$. Further studies showed that substituted cyclooctatetraenes C_8H_7R (R is Me, Ph, or $SiMe_3$), or the cyclooctatrienes C_8H_8-$(SiMe_3)_2$ and $C_8H_7(SiMe_3)_3$, react with either $[Ru_3(CO)_{12}]$ or compound **54a** in heptane or octane at reflux to give, among other

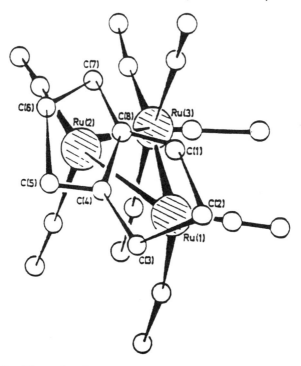

Figure 2. The molecular structure of the pentalene–ruthenium complex [Ru$_3$(CO)$_8$(η^8-C$_8$H$_6$)]. (Reproduced with permission from reference 130e. Copyright 1974. See also references 130b and 130c.)

products, a family of [Ru$_3$(CO)$_8$(pentalene)] complexes.[130] Formation of trimethylsilyl-substituted pentalenes was favored, with yields in some reactions as high as 30 to 40%. Moreover, it was found that [Ru$_3$(CO)$_8$(η^8-C$_8$H$_6$)] and species containing symmetrically substituted pentalenes exist as two isomers. These isomers arise because the hydrocarbon can either edge- or face-bond the Ru$_3$ triangle. This bonding was confirmed by several X-ray diffraction studies, carried out by Judith Howard and Peter Woodward and their students, including the structure determination of both edge- and face-bonded isomers of the tris-(trimethylsilyl)pentalene complex [Ru$_3$(CO)$_8${C$_8$H$_3$(SiMe$_3$)$_3$-1,3,5}] (the 1,3,5-locants indicate that the SiMe$_3$ groups are substituted at the 1, 3, and 5 positions of C$_8$H$_3$). This compound forms crystals of two types. In one type the pentalene ligand edge bridges the Ru$_3$ triangle so that the plane defined by the metal atoms is at ~50° to the plane of the eight carbon atoms, as it is in

*A 1973 visit to Spain: in the Pyrénées with Rafael and Sonja Usón
and Judy Stone. Under the leadership of Professor Usón of Zaragoza,
research in organometallic chemistry has developed strongly in Spain
at several centers. Many postdoctoral co-workers from Spain studied in
my Bristol laboratory from 1973 through 1990.*

$[Ru_3(CO)_8(\eta^8\text{-}C_8H_6)]$ (Figure 2). In the crystal of the other isomer, the $C_8H_3(SiMe_3)_3$-1,3,5 molecule lies symmetrically parallel
to the Ru_3 triangle, bound to all three metal atoms. This
geometry necessitates a rearrangement of the carbonyl ligands,
leading to the presence of two $Ru(CO)_3$ and one $Ru(CO)_2$ group,
in contrast with the other isomer, which has one $Ru(CO)_4$ and
two $Ru(CO)_2$ groups. Amusingly, but only in retrospect, the
very interesting diffraction data obtained for the two isomers
were not analyzed until several weeks after the they were
obtained. After this experience I have become more persistent in
seeking results from colleagues in crystallography, asking frequently whether a structure has been solved following data collection or whether the raw data remains unexamined in the com-
puter.

Four of the 21 postdoctoral Spanish colleagues who worked in my group at Bristol. Left to right: Professors José Vicente (Murcia), Victor Riera (Oviedo), me, Juan Forniés (Zaragoza), and Dr. Miguel Ciriano (Zaragoza). All came from Professor Rafael Usón's group at Zaragoza.

Complexes containing unsymmetrically substituted pentalenes exist exclusively as edge-bonded isomers and do not undergo dynamic behavior. In contrast, $[Ru_3(CO)_8(\eta^8-C_8H_6)]$ and analogs with symmetrically substituted pentalene ligands were shown by ^1H NMR studies to be fluxional, with the predominant edge-bonded isomer undergoing the oscillatory process (ii) depicted in Scheme IIIa for $[Ru_3(CO)_8\{\eta^8-C_8H_5(SiMe_3)-2\}]$. The oscillation may be viewed either as a pendulum-like swing of the $Ru(CO)_4$ group with respect to the hydrocarbon, or as a movement of the pentalene about one edge of the Ru_3 triangle. The isomerization (process i of Scheme IIIa) is remarkable in requiring the shift of a hydrocarbon ligand from the edge of a metal cluster to a face and vice versa, with simul-

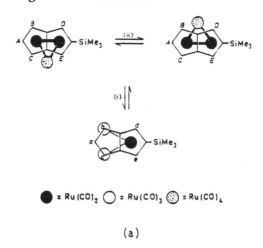

(a)

(b)

*Scheme III. The dynamic behavior of pentalenetriruthenium com-
plexes. (a) The fluxionality of the symmetrical complex [Ru$_3$(CO)$_8$-
{η^8-C$_8$H$_5$SiMe$_3$-2}] involves an oscillatory process ii and an edge-to-
face movement of the pentalene ligand i. (b) Processes i and ii for
[Ru$_3$(CO)$_8$(η^8-C$_8$H$_6$)] allow migration of the hydrocarbon group over
all edges and faces of the metal cluster. (Reproduced with permission
from reference 130b. Copyright 1979.)*

taneous carbonyl migration between ruthenium atoms. For the compound $[Ru_3(CO)_8(\eta^8-C_8H_6)]$ a combination of both fluxional processes allows migration of the pentalene over all edges and both faces of the metal cluster (Scheme IIIb). Migration of a hydrocarbon group in this manner had never previously been demonstrated, and the mobility of a polyolefin or aromatic compound on a metal surface in catalysis between "steps", "kinks", and "terraces" may be modeled by the behavior described here.

Thus was completed a very satisfying study[132] spread over several years, involving the interesting organic ligand pentalene, that began as a result of the fortuitous discovery of the diruthenium complex 63b. In addition to contributions by Victor Riera, important discoveries were made by postdoctoral assistant Ron McKinney, who came to us from Herb Kaesz's laboratory at UCLA and who is now at DuPont, and by Bristol graduate students Tony Brookes, Jenny Burt, Julie Edwards (later Knox), Paul Harris, Richard Phillips, Barry Sosinsky, Anne Szary, and Mark Winter, now a faculty member at Sheffield University.

From Metallacarboranes to "Platinum with Wings"

In December of 1970, John L. Spencer from Otago University, New Zealand, joined my group at Bristol as a postdoctoral fellow and set about trying to prepare a platinum species "$[Pt(CNBu^t)_2]$" by treating $[Pt(\eta^3-C_3H_5)(\eta^5-C_5H_5)]$ with $CNBu^t$. He tried this synthesis because the palladium analog "$[Pd(CNBu^t)_2]$", used in the preparation of compounds such as 45, had been obtained by treating the complex $[Pd(\eta^3-C_3H_5)(\eta^5-C_5H_5)]$ with $CNBu^t$. However, the reaction between $[Pt(\eta^3-C_3H_5)(\eta^5-C_5H_5)]$ and $CNBu^t$ did not yield "$[Pt(CNBu^t)_2]$", so its projected reactions with fluorocarbons could not be studied.

This failure accelerated my waning enthusiasm for fluorocarbon–metal chemistry, and I drew John Spencer's attention to the chemistry of metallacarboranes. I had been greatly impressed by the superb work of Fred Hawthorne's group in this area, and I had retained a strong interest in boron chemistry following my postdoctoral period with Anton Burg. I had met Fred in the fall of 1960 when he came to Harvard to teach a course in physical organic chemistry, and we have been friends since that time. While at Harvard he was on leave from the Redstone Arsenal Division of the Rohm and Haas Company, at a time when the company had a strong research program directed at the use of boron compounds as rocket fuels. Later in 1965, after Hawthorne had left industry for academia, he and his co-workers[133] reported the synthesis of the anionic iron complex $[Fe(\eta^5-C_2B_9H_{11})_2]^{2-}$ (**65**), having recognized that the open pentagonal face of the *nido*-cage fragment $C_2B_9H_{11}^{2-}$ could function as a ligand like $C_5H_5^-$; today we would describe these species as being isolobal, a term defined more precisely later in this book.[68] Subsequently, many complexes of other metals containing the ligands η^5-7,8- or η^5-7,9-$C_2B_9H_9R'_2$ (R' is H or Me) were prepared by Hawthorne's group, and a rich new area of organometallic chemistry was born.

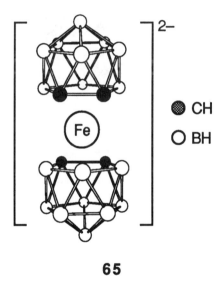

● CH
○ BH

65

I suggested to John that he prepare the compound $[Pt(cod)(\eta^5-7,8-C_2B_9H_{11})]$, and some related compounds, in order to ascertain how the $C_2B_9H_{11}^{2-}$ group might influence the reactivity of other ligands present in the complexes. This line of endeavor proved unproductive except for one discovery. In order to prepare the carborane complexes from metal chlorides, John employed $Tl[TlC_2B_9H_{11}]$ as a source of $C_2B_9H_9^{2-}$, recognizing that in the syntheses the chloride ion would be removed as insoluble TlCl, thereby providing a driving force for reaction.

Diana and Fred Hawthorne visited Baylor University in the spring of 1992, when Fred gave the Gooch–Stephens lectures.

We communicated[134] the preparation of $Tl[TlC_2B_9H_{11}]$ in August 1972, and this reagent and its C-methyl derivatives were later used extensively by Hawthorne's group,[135] and also by Malcolm Wallbridge and his co-workers.[136] Indeed, even today these species first identified by Spencer are presenting important structural problems, as interest focuses on metallacarboranes containing the heavier group 13 metals.[135]

In 1970 Dunks and Hawthorne[137] had described a novel strategy for preparing metallacarboranes. This procedure was called the "polyhedral expansion" method, a two-step synthesis that depended upon reducing a neutral *closo*-carborane to give a *nido*-carborane dianion, which was then treated with a transition metal salt to reform a *closo* structure containing the metal and having one extra vertex. The particular article[138] that caught my eye appeared in May 1971; it described the polyhedral expansion of the 12-vertex carborane $C_2B_{10}H_{12}$ to give a 13-vertex cobaltacarborane:

$$C_2B_{10}H_{12} \xrightarrow{Na^+C_{10}H_8^-} C_2B_{10}H_{12}^{2-} \xrightarrow[\substack{C_5H_5^- \\ [O]}]{Co(II)} [Co(\eta^6\text{-}C_2B_{10}H_{12})(\eta^5\text{-}C_5H_5)]$$

These results suggested to me that some of the expanding range of zero-valent metal complexes of Ni, Pd, or Pt might undergo a one-step oxidative insertion into *closo*-carboranes. I anticipated that the d^{10} metal compounds would function as the source of the two electrons required to open the carborane cage. The latter would then be closed again by capture of the d^8 metal–ligand fragment resulting from the electron transfer. In July 1971, John Spencer tested this idea using $[Ni(PEt_3)_4]$ and $C_2B_{10}H_{12}$. No reaction occurred between these two reagents, but with hindsight this negative result was explicable in terms of the high stability of the icosahedral 12-vertex structure of the carborane chosen. Thus not surprisingly Ni^0 is not as effective a reducing agent for opening *closo*-carborane cages for metal atom insertion as is Hawthorne's use of sodium naphthalenide.

Because of a protracted holiday in Scandinavia, John's next experiments were not carried out until October 1971. In the interim, George Parshall, Earl Muetterties, and their productive colleagues[139] at DuPont had prepared $[Pt(PEt_3)_3]$ by treating $[Pt(\eta^3\text{-}B_3H_7)(PEt_3)_2]$ with PEt_3, a result that led John Spencer to synthesize the compounds $[Pt(PEt_3)_n]$ (n is 3 or 4) from the reaction between $[Pt(\eta^3\text{-}C_3H_5)(\eta^5\text{-}C_5H_5)]$ and an excess of PEt_3. A second attempt was then made at direct insertion of a metal–ligand fragment into a carborane by using $[Pt(PEt_3)_3]$ and $MeC_2B_{10}H_{11}$. Again nothing came of this reaction, and I would have abandoned this route to metallacarboranes had not John Spencer appreciated that we should be employing carboranes with fewer than 12 vertices. Hence, in the polyhedral expansion process we would aim to attain the very stable icosahedral structure rather than attempt to depart from it. Thus Spencer was led to treat the 11-vertex $MeC_2B_9H_{10}$ with $[Pt(PEt_3)_3]$ and so obtained a quantitative yield of pale lemon-yellow crystals of

$[Pt(PEt_3)_2(\eta^5\text{-}MeC_2B_9H_{10})]$. It was then discovered[134] that $Me_2C_2B_9H_9$ afforded a range of icosahedral metallacarboranes containing Ni, Pd, or Pt in reactions with the reagents $[Ni(cod)_2]$, $[Pd(CNBu^t)_2]$, $[Pt(PMe_2Ph)_3]$, and so forth. Moreover, it was later found, with many fruitful ideas suggested by Michael Green, that carboranes with 6–10 vertices would combine with d^{10} metal complexes to afford a variety of metallacarboranes, some including two-metal atoms.[140] Many necessary X-ray diffraction studies were carried out by Dr. Alan Welch, with synthetic studies by Drs. Geoff Barker and by Maria Pilar Garcia from Zaragoza.

I have given an account elsewhere[141] of some of our early results on metallacarboranes, and since the topic is perhaps more "inorganic" than "organometallic" I shall not review it further here. However, there is a moral to the story of John Spencer's success in opening up this area for us. As mentioned, initially there were many negative results. When entering a new field negative results are not unusual, and in these circumstances a supervisor has several options. My philosophy has been either to abandon the project when the going gets very difficult or hope that a young co-worker will have the insight and perseverance to find a route to success, as did John Spencer, before becoming too frustrated. The former choice is depressing for the supervisor if he is convinced that his ideas are sound in principle; moreover, judgement is required as to when to give up. However, this tactic is much preferable to the alternative that one hears is practiced in some groups; another co-worker being assigned the same project to see if he or she can make it work. This is a certain way to create disharmony and bad feeling, unless the first co-worker who tried out the idea has long departed from the scene and one does not hide the earlier unsuccessful experiments from the new young colleague.

Almost as a sideline to the metallacarborane research, John Spencer continued to work on the synthesis of Pt^0 compounds. These studies were in part encouraged by the successful use of $[Pt(PEt_3)_3]$ to prepare the novel dihaptoareneplatinum complex **40**, discussed earlier. A useful preparation of $[Pt(stilbene)(PMe_3)_2]$ was developed by reducing cis-$[PtCl_2(PMe_3)_2]$ with $Na[AlH_2(OCH_2CH_2OMe)_2]$ in the presence of stilbene. However, we were aware of the existence of the potentially interesting compound $[Pt(cod)_2]$, and another co-worker, David Empsall from Cambridge, had spent some time in 1971 trying to repeat the reported synthesis. Müller and Göser[142] had prepared [Pt(cod)$_2$] by first converting [PtCl$_2$(cod)] to [PtPr$^i_2$(cod)], and then in a second reductive step the latter had been irradiated with UV light in the presence of cycloocta-1,5-diene to yield the desired product. In our hands, we found this method difficult to repeat.

Our thoughts became focused on [Pt(cod)$_2$] because of the renaissance of organonickel chemistry stemming from the discovery of "pure alkene" complexes of nickel by Wilke and his co-workers[111,143] in the early 1960s. These very reactive compounds may be divided into two classes: tris(olefin) complexes, for example, [Ni(*t,t,t*-cyclododeca-1,5,9-triene)], or species in which four C=C groups ligate the nickel, for example, [Ni(cod)$_2$]. Moreover, as described earlier, we had gained experience with these compounds as reagents in our syntheses of fluorocarbon–nickel compounds.

It was in March 1974 that John Spencer carried out the first successful preparation of [Pt(cod)$_2$] by reducing [PtCl$_2$(cod)] with Li$_2$C$_8$H$_8$ in the presence of cod.[144,145] We chose this novel reducing agent because it had the advantage that the cyclooctatetraene released after electron transfer from C$_8$H$_8^{2-}$ does not form complexes with Pt^0 in the presence of an excess of cod.[146] The first reaction of [Pt(cod)$_2$] studied was that with CNBu^t, and the product obtained in quantitative yield was the triplatinum compound [Pt$_3$(CNBu^t)$_6$] (66).[147] The structure was established

66

by X-ray diffraction by Judith Howard, who gave me the results in a lighter moment at a taverna on Corfu when we were attending a conference! One could not imagine a more pleasant venue to hear a new result. The complex $[Pt_3(CNBu^t)_6]$ proved to be an important molecule in its own right, and it stimulated Michael Green and me to have our co-workers Geoff Barker and Jean-Marie Bassett prepare the species $[Fe(CNR)_5]$ and $[Fe_2(CNR)_9]$, analogs of $[Fe(CO)_5]$ and $[Fe_2(CO)_9]$, respectively.

After the breakthrough with the preparation of $[Pt(cod)_2]$, progress was very rapid. We were aware that if we did not exploit our discoveries quickly, others would surely do so.

Judith Howard and John Spencer relax at Corfu in June 1974 while attending a NATO Advanced Study Institute. Judith Howard carried out numerous X-ray diffraction studies on complexes made by my group over a period of 25 years, and as described in the text, John Spencer made important discoveries in many areas, including fluoro-organometal compounds, metallacarboranes, and alkene–platinum complexes.

Organometallic chemistry is a very competitive area, and just as an actor is judged by his most recent performance, so organometallic chemists are judged by their most recent work. I have never had difficulty in persuading my co-workers to concentrate almost solely on laboratory work when the occasion arises. During the summer and fall of 1974 bis(cycloocta-1,5-diene)platinum was used as a precursor for the synthesis of $[Pt(C_2H_4)_3]$, $[Pt(C_2H_4)_2(C_2F_4)]$, $[Pt(C_2H_4)_2(PR_3)]$, and $[Pt(PR_3)_2]$ [PR_3 is $P(Bu^t_2\text{-}Me)_3$ or $P(cyclo\text{-}C_6H_{11})_3$]. Preparations of the strained alkene complexes $[Pt(norbornene)_3]$ and $[Pt(trans\text{-}cyclooctene)_3]$ were also devised in this period. The synthesis of $[Pt(C_2H_4)_3]$ by treating $[Pt(cod)_2]$ with C_2H_4 came about in the following way. The Mülheim school[148] had reported the very unstable complex $[Ni(C_2H_4)_3]$, obtaining this species by treating $[Ni(t,t,t\text{-cyclo-}$ dodeca-1,5,9-triene)]$ with C_2H_4. The Mülheim chemists had also observed that $[Ni(cod)_2]$ with C_2H_4 afforded $[Ni(C_2H_4)_3]$, *but the reaction did not go to completion since an equilibrium existed between these 18- and 16-electron nickel species.* Since platinum prefers a 16-valence-electron shell, it seemed likely to us that if an equilibrium existed between cod and C_2H_4 at a platinum center, then in hydrocarbon solvents the equilibrium would favor $[Pt(C_2H_4)_3]$ rather than $[Pt(cod)_2]$. This preference proved to be so, as ethylene readily displaces cod from the latter. Tris(ethylene)-platinum forms white crystals that can be purified by sublimation at 20 °C in the presence of ethylene. In the absence of ethylene the solid complex decomposes to platinum metal. Its chemistry can be conveniently explored by generating hydrocarbon solutions of the compound in situ from $[Pt(cod)_2]$. Moreover, treatment of these solutions with one or two equivalents of PR_3 provides a convenient route to the compounds $[Pt(C_2H_4)_2(PR_3)]$ and $[Pt(C_2H_4)(PR_3)_2]$, respectively, which are excellent sources of the fragments $Pt(PR_3)$ and $Pt(PR_3)_2$ for further synthesis.

The chemistry of the species $[Pt(C_2H_4)_3]$, $[Pt(cod)_2]$, and $[Pt(C_2H_4)_2(PR_3)]$ proved to be very extensive, and I have given an account of the early work elsewhere.[149] Major contributions were made by Geoff Barker, Neil Boag, Miguel Ciriano, David Grove, Juan Forniés, and Antonio Laguna (Juan Forniés and Antonio Laguna are now professors of chemistry at the University of Zaragoza, Spain). Whereas $[Pt(cod)_2]$ often affords products containing a Pt(cod) fragment, $[Pt(C_2H_4)_3]$ is a source of

Judith Howard shows John Spencer and me models of $[Pt(C_2H_4)_3]$, the structure of which she had determined (Chem. Eng. News, **1975**, January 20).

"naked platinum". Some appreciation of the value of these compounds may be gained by considering the situation prior to their discovery. Precursors for exploring zero-valent platinum chemistry were limited to the phosphine complexes $[Pt(PR_3)_n]$ (n is 3 or 4) or $[Pt(alkene)(PR_3)_2]$. Although important reagents, these compounds have the disadvantage of containing strongly ligating PR_3 groups that occupy coordination sites and hence reduce the reactivity of the metal center. For the development of the organic chemistry of nickel the situation was much more favorable. The very labile species $[Ni(CO)_4]$ had been available for decades, and the isolation of the "naked nickel" complexes $[Ni(cod)_2]$ and $[Ni(t,t,t\text{-cyclododeca-1,5,9-triene})]$ by the Mülheim school[143] had led to the discovery of numerous organic syntheses based on nickel. The platinum carbonyl $[Pt(CO)_4]$ exists only as a transient species in a matrix at very low temperatures and is thus useless as a chemical reagent. Mond's isolation of $[Ni(CO)_4]$ in 1890 prompted Lord Kelvin colorfully to remark that this discovery "had given wings to metals". The isolation of $[Pt(C_2H_4)_3]$ continued this tradition, hence the title of this section. The structure of the compound was established by Judith Howard with both X-ray and neutron diffraction techniques[150] and was important in revealing a trigonal-planar structure with all six alkene carbon atoms lying in the coordination plane, rather than the alternative with a trigonal-upright arrangement where the C=C double bonds are perpendicular to the coordination plane. However, the first structural studies[151] were on the two compounds $[Pt(C_2H_4)_2(C_2F_4)]$ and $[Pt(norbornene)_3]$, and the results appeared shortly after the publication of a paper by Rösch and Hoffmann[152] that correctly predicted the ground-state structures for d^{10} molecules of the type $[M(alkene)_3]$. Our discovery of $[Pt(C_2H_4)_3]$ and related molecules merited mention in *Chemical & Engineering News*, and a photograph that accompanied the article is shown on page 106.

Roald Hoffmann's prediction of the correct structure of the species $[M(alkene)_3]$ provides a beautiful example of his many seminal contributions to chemistry. We were not aware of the work described in reference 152 until after the paper had appeared, because at the time I was not on Roald Hoffman's mailing list for preprints. Later Roald generously sent me numerous preprints of his work prior to publication and we, as must have many others, benefited from his unique insight and breadth of vision (he is the only person so far to have received both the American Chemical Society's A. C. Cope Award in Organic Chemistry and the Society's Award for Research in Inorganic Chemistry). However, in the domain of organometallic chemistry those engaged in synthesis have usually been far ahead of predictions made by theoretical chemists. Indeed, it has been remarkable what exciting experimental discoveries have been made by preparative workers relying on their intuition and an ability to count up to 16 or 18 valence electrons at the metal center! Indeed, predictions from theoretical chemists can sometimes impede preparative work, as in Wilkinson's story about how he was put off from trying to make $[Cr(C_6H_6)_2]$.* However, these strictures do not apply to the work of Hoffmann, who has used theory to point the way ahead for preparative chemists, as in some of the chemistry described later.

---

* *See* G. Wilkinson, *J. Organomet. Chem.* **1975**, *100*, 273. However, much credit must be given to the theoretical chemists Longuet-Higgins and Orgel for predicting, on the basis of MO calculations, that a square planar cyclobutadiene molecule would be stabilized by coordination to a transition metal, *See* reference 64. Moreover, the concept of the three-center two-electron bond, mentioned at several places in this book, was first proposed by Longuet-Higgins, *see J. Chim. Phys.* **1949**, *46*, 268.

Tris(ethylene)platinum

When the complexes [Pt(cod)$_2$], [Pt(C$_2$H$_4$)$_3$], and [Pt(C$_2$H$_4$)$_2$(PR$_3$)]
became available to us in 1974 there was no shortage of interest-
ing reactions to investigate. By chance also in 1974 volume 1 of
The Organic Chemistry of Nickel appeared, a book[143] that superbly
summarized in detail the results of the Mülheim group. Since I
was one of the editors of the series of monographs in which the
volume appeared I received an advanced copy, which I gave to
Michael Green for bedside reading. Michael was quick to suggest a number of experiments for the platinum compounds based
on what had been reported previously for the "naked nickel"
complexes. Progress was rapid in the hands of the several good
co-workers mentioned, and also by Teresa Chicote, her husband
José Vicente (now a professor of chemistry at the University of
Murcia, Spain), and Gary Scholes. Consequently, the boundaries
of organoplatinum chemistry were greatly extended over a relatively short period.[149] Not surprisingly the reactivity patterns
for the new platinum–alkene complexes were different from
those of their nickel analogs. Thus, whereas [Ni(cod)$_2$] cyclo-
trimerizes butadiene to cyclododeca-1,5,9-triene, the platinum
compound [Pt(cod)$_2$] affords the stable 2,5-divinylplatinacyclo-
pentane complex **67a**. Species with similar structures have been
invoked as intermediates in the nickel-catalyzed reaction. Treatment of butadiene with [Pt(C$_2$H$_4$)$_2$(PMe$_3$)] affords the
octadienediyl(trimethylphosphine)platinum compound **68**, which

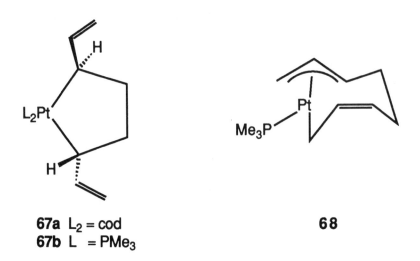

67a L$_2$ = cod   **68**
67b L = PMe$_3$

with an equivalent of PMe_3 yields **67b**.[153] Moreover, treatment of complex **67a** with one equivalent of PMe_3 gives compound **68**. Butadiene is catalytically cyclodimerized by $[Ni(cod)_2]$ in the presence of a tertiary phosphine, and open-chain C_8 complexes akin to **68** play a key role. Thus, although with platinum the reactions are not catalytic, they have allowed the isolation of model intermediates. This fact reflects the more stable carbon–metal bonds formed by platinum compared with nickel. Hence, reductive elimination processes leading to C–C bond formation and catalysis are not observed. However, later studies[154] showed that the compounds $[M(cod)_2]$ (M is Pd or Pt) would catalyze the telomerization of butadiene in the presence of secondary amines.

My graduate students Neil Boag and David Grove showed that alkynes react readily with $[Pt(cod)_2]$ or $[Pt(C_2H_4)_3]$ to give a range of products (Chart I, **A–G**), the nature of which depends on the alkyne used and the stoichiometry employed.[155] David Grove was a Bristol graduate, but Neil Boag, now teaching at Salford University, had obtained his bachelor's degree with first class honors from Imperial College. Having decided that he did not wish to continue his studies at IC, fortunately for me he decided to join my group, after he had visited several other research centers.

It was salutary for me to learn some years later the reasoning behind his decision, coming to light after I had recommended him for academic positions. When we first met he observed that I was wearing a sweater with a large hole under the armpit, and decided that my "homely" appearance indicated a relaxed friendly environment in which he could work. The lesson to be learned, apart from the one that I should have neglected my dress even more while at Bristol in order to attract good co-workers from other universities, is that students select Ph.D. supervisors for a variety of often curious reasons, and not necessarily on the basis of erudite accounts of research currently under way in the laboratory. Many candidates are more impressed by finding themselves in an atmosphere in which they are treated as equals than by having to listen in a pupil–teacher relationship to the very latest results in a field with which they are unfamiliar. Moreover, meeting a potential co-worker at an informal occasion can also be useful. Thus my first meeting with one of my most talented postdoctoral associ-

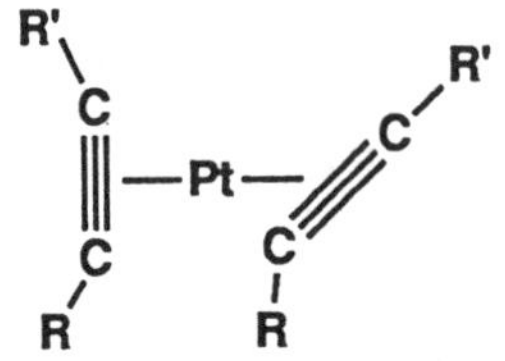

A

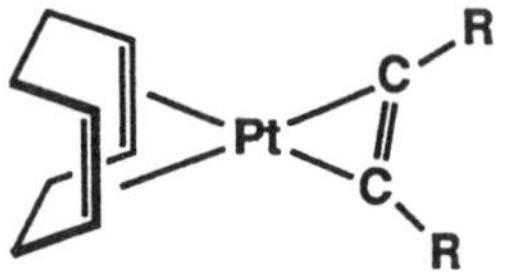

B

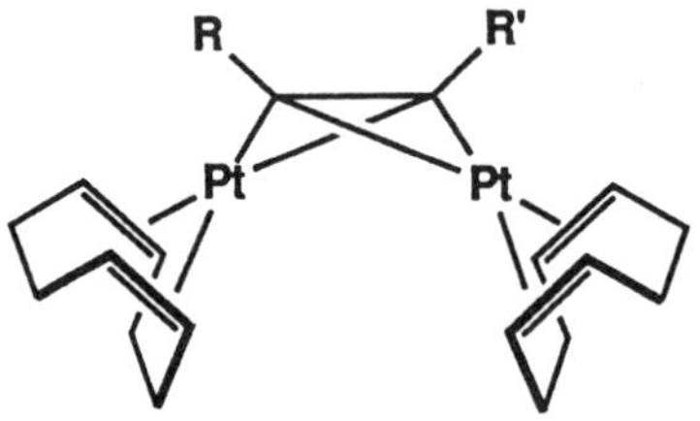

C

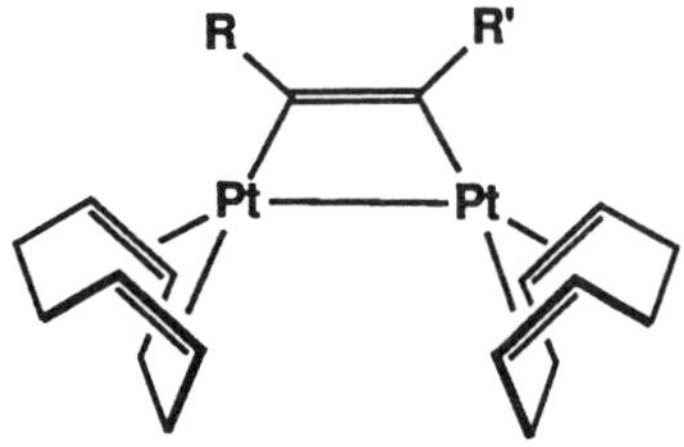

D

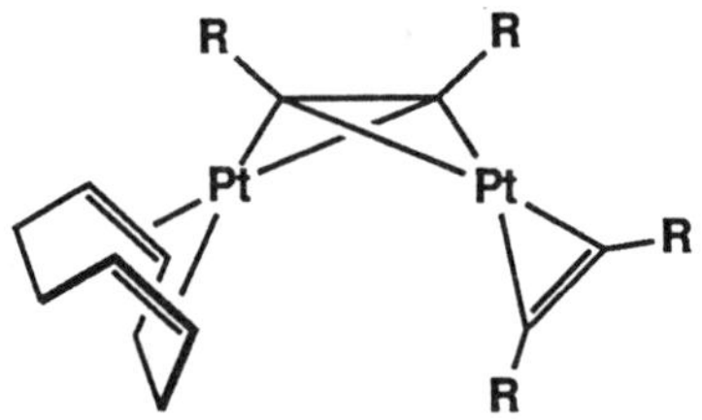

E

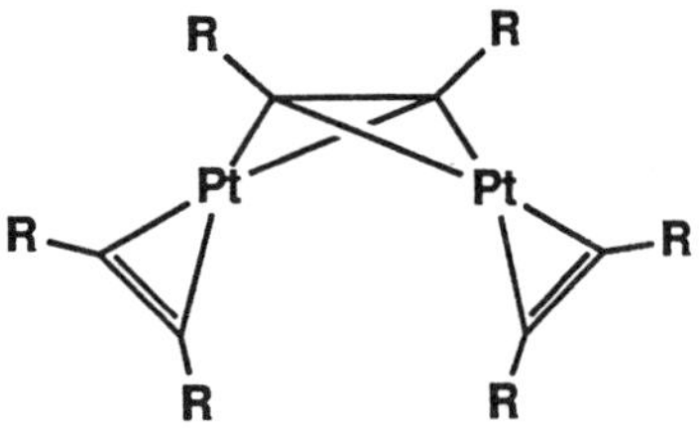

F

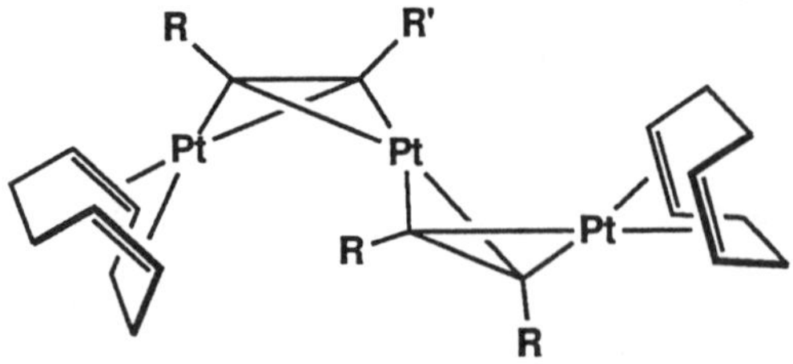

G

Chart I. Structures of some alkyneplatinum complexes prepared from [Pt(cod)$_2$] or [Pt(C_2H_4)$_3$].

ates, Dr. Michel Pfeffer of the Centre National de la Recherche Scientifique (CNRS), took place in the casino at Venice, an occasion well worth the loss of a little money.

Reverting to the extensive chemistry of low-valent platinum–alkyne complexes, in structures **C, E, F,** and **G** in Chart I, the metal atoms and the ligating carbon atoms of the bridging alkynes adopt a dimetallatetrahedrane configuration, although the Pt-to-Pt separations (~2.90 Å) suggest little or no direct metal–metal bonding. The disposition of the two orthogonal orbitals of the bridging alkynes probably determines the stereochemistry of the Pt(μ-C$_2$)Pt fragments. It is likely that species with more than three platinum atoms can be prepared by stepwise addition of alkynes and [Pt(C$_2$H$_4$)$_3$] to complexes of type **G,** but this possibility remains to be explored. It is noteworthy that **C** and **D** are isomers, with the alkyne in the latter lying parallel to the metal–metal bond so as to form a dimetallacyclobutene ring structure.

The complexes [Pt(C$_2$H$_4$)$_2$(PR$_3$)] have a rich chemistry. With the silanes SiHR′$_3$ and SiH$_2$R′$_2$ they afford diplatinum compounds [Pt$_2$(μ-H)$_2$(SiR′$_3$)$_2$(PR$_3$)$_2$] (**69**) and [Pt$_2$(μ-SiHR′$_2$)$_2$(PR$_3$)$_2$] (**70**),[156] respectively. Spectroscopic and other evidence suggests that in the species **69** and **70** the hydrido ligands participate in multicenter bridge-bonding between platinum and silicon. This type of "agostic" Si–H $\longrightarrow$ metal interaction was discovered much earlier by Bill Graham and his co-workers[157] in important studies on the compounds [MnH(SiPh$_3$)(CO)$_2$(η^5-C$_5$H$_5$)] and [Re$_2$H$_2$(μ-SiMe$_2$)$_2$(CO)$_8$]. The compounds **69** are excellent catalysts for hydrosilylation.[158]

Among other early studies[149] with [Pt(C$_2$H$_4$)$_3$] and [Pt(cod)$_2$] are reactions with unsaturated organofluorine compounds, deserving of special mention. Several interesting metallacycles were characterized, as illustrated by compounds **49, 50,**

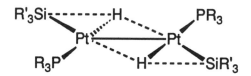

69

70

and **52** mentioned earlier. For reasons described later, it was, however, a reaction between $CF_3CF=CF_2$ and $[Pt(cod)_2]$, studied by Antonio Laguna, that was of the greatest significance for future work. The product was the diplatinum compound **71a**,[159] arising from a fluorine migration to give the bridging bis(trifluoromethyl)methylene ligand. By treating $[Pd(cod)_2]$, first prepared in our laboratory,[144] with $CF_3CF=CF_2$, we were able to isolate as an intermediate the alkenepalladium complex

Bill Graham (left) and Herb Kaesz at Bristol in 1987 standing by one of the nails outside the Corn Exchange. The nails (another may be seen in the background) were once used to settle trading deals between farmers and merchants. Hence the phrase "cash on the nail".

71a Pt
71b Pd

$[Pd\{\eta^2\text{-}CF_2{=}CF(CF_3)\}(cod)]$ and demonstrate that this species rearranges in Et_2O to yield **71b**, the palladium analog of **71a**. To explain the pathway from the fluoroalkene complexes to the compounds **71**, Michael Green suggested that the former underwent fluoride ion migration to give the ylide species $(cod)M^+{-}^-C(CF_3)_2$. The latter were then presumed to capture M(cod) (M is Pd or Pt) fragments to yield the final products.

Travels

At this juncture it is convenient to interrupt the survey of the chemistry with which I have been associated to write about my good fortune in being able to visit many different parts of the world as a result of being active in research. Although I had first begun attending scientific meetings when I was working at Harvard, and indeed had visited Australia for the first time in 1962, it was not until I attended the International Conference on Coordination Chemistry held in Haifa and Jerusalem in September 1968 that I fully appreciated that my activities in research could be parlayed into enjoyable travel experiences. For example, I have a very happy memory of walking from the Dead Sea to the top of Masada with Joseph Chatt and Ronald Nyholm, starting at 6 a.m. before the temperature climbed to 40 °C. I believe that since that time a chair lift has been installed for visitors. Both Joe Chatt and Ron Nyholm led the renaissance of inorganic chemistry in Great Britain, and I was grateful for their friendship and support over many years. Very sadly, Ron Nyholm was killed in an automobile accident in 1971 at the age of 54. Had he lived, his tireless advocacy of the claims of chemistry and his truly remarkable personal qualities, coupled with his influential position in British science, might well have served to negate some of the policies so detrimental to chemistry adopted in more recent times and referred to later in this book.

In 1969 the Fourth International Conference on Organometallic Chemistry was held at Bristol under my chairmanship. With more than 500 active participants, it was the largest scientific meeting held at the university up until that time. No expensive "professional conference organizers" were hired to make the arrangements, as subsequently became all too common for such meetings. By dint of much hard work, accomplished over a period of many months, we were able to make all the arrangements in-house, thereby eliminating much administrative expense and keeping the costs for participants low. We were pleased that our efforts enabled many graduate students from universities in Britain, France, Germany, and the other European countries to attend.

Many persons from Eastern Europe indicated their intention to attend the conference, and in that era this created special problems that many older readers will appreciate from their own personal experiences. We received abstracts of papers for oral presentation from several chemists in the former Soviet Union. Aware of the probability that these persons might not be allowed to come to Bristol, we scheduled their papers for delivery at the ends of sessions so that if they did not appear there would be minimal disruption of the scientific program. It required no high level of initiative on our part to do this, nor to anticipate that a proportion of those who managed to reach Bristol would not in practice be those who had registered or had submitted the papers in the first place. Nevertheless, those of my colleagues involved in the organization of the Conference, of whom Eddie Abel, Michael Bruce, Peter Goggin, and Peter Timms deserve special credit, were able to rise to unanticipated emergencies. Participants from the Soviet Union were unable to send funds in advance for registration or for their accommodation, thereby making planning difficult.

With the conference due to begin with a reception on a Sunday evening we received no information about the Russians until the previous day, when Eddie Abel, as conference secretary, took a telephone call from a London office of the travel agent Thomas Cook informing him that a bus containing the USSR delegation would be heading our way from Heathrow early next morning and inquiring where they should be depos-

In 1969 the 4th International Conference on Organometallic Chemistry was held in Bristol with more than 500 participants. The photograph shows from left to right Dick Pilling, Michael Bruce, and me preparing conference registration material.

ited in Bristol. Ever resourceful, Eddie found an obscure hotel located in a quiet area nearby and booked all its rooms for the week for the participants from the former Soviet Union. Although this prompt action solved our problem, unfortunately it made life easier for two or three members of the group who were observed during the period of the conference to be much less interested in the chemistry presented than in the whereabouts of their colleagues at all times. Fortunately, it did not become necessary to use the emergency telephone number given to us by the U.K. Special Branch prior to the meeting for our use should anyone from one of the Eastern countries wish to join my group on a permanent basis. Another aspect of the Cold War that reared its head during the conference was a certain amount of abuse directed at Peter Timms by the East Germans for his inadvertently having listed them with their West German colleagues, a procedure likely to get them into trouble on their return home. Happily such incidents belong to the past, and hopefully within a few years will probably be regarded by younger colleagues as unbelievable tales of older colleagues in chemistry.

Although during my time at Bristol I organized other meetings, including in 1981 the First International Conference on the Platinum Group Metals, after 1969 I became more enamored with attending conferences elsewhere, arranged by others, and responding to numerous invitations to give talks on my research in other countries. To my wife's chagrin I never took a sabbatical leave during my 27 years at Bristol. However, hardly a year went by when I did not spend a week or a month visiting universities in North America. Most British universities do not have formal study leave programs, although it is possible to be absent for a protracted period, provided one has arranged matters so that teaching and other commitments are met at other times during the academic year. However, I never wished to be away from my research group for a long period. I am a firm believer in the need for research supervisors to be on hand to remove as far as possible any real or imagined impediments placed in the way of the work being done by their young colleagues. In the British system, which is geared to students completing their Ph.D. studies in 3 years, protracted periods of absence on the part of a supervisor are detrimental. Relatively short absences are a different matter, and it occasionally happened that some interesting discoveries were made while I was away; when this happened I was not allowed to forget it by my younger co-workers for several months.

Although I was never absent from Bristol for very long periods, apart from a 6-month visit to Princeton in 1967, I traveled frequently and continue to do so; for want of something better I have even listed "travel" as my hobby in *Who's Who*. Many enjoyable visits were made to Spain, a country from which several of my co-workers originated. These visits followed Victor Riera's profitable stay in my group in 1973–1974 and my development of a friendship with his mentor, Professor Rafael Usón. I spent two U.K. summer periods in Australia and New Zealand sampling the hospitality of these countries. Under various international programs of the Royal Society, I made visits lasting 2 or 3 weeks to Finland, Hungary, India, Israel, Japan, Poland, Sweden, and what was the Soviet Union. On a visit to the latter in 1978 it was a great pleasure for Judy and me to be entertained by Professor Aleksandr Nesmeyanov and Madame Marina Nesmeyanov at their country house in the woods some 30 miles from Moscow. Early in the spring, with

Hard-working organometallic chemists at the Symposium on Relations between Homogeneous and Heterogeneous Catalysis held at Asilomar, California, October 1983. Left to right: Sheldon Shore (Ohio State), me, Duward Shriver (Northwestern), Herbert Kaesz (UCLA, conference organizer), and Robin Whyman (I.C.I., United Kingdom).

snow lying in the surrounding woods, it was a memorable experience. During his long life Nesmeyanov was associated with extensive and internationally acclaimed studies on organometallic chemistry. The Index of *Comprehensive Organometallic Chemistry* carries more than 700 references to his work, which he preferred to define as "organoelement chemistry" in order to widen its scope. He was a great administrator as well as a scientist. A large university complex was built on the Lenin Hills during the period (1948–1951) when he was rector of Moscow University, and for 10 years he was president of the USSR Academy of Sciences, which now no longer exists, having been merged into the new Russian Academy of Sciences. In 1961 Nesmeyanov advanced the idea of food production by industrial methods avoiding agriculture, and with some of his many collaborators at the Institute of Organoelement Compounds in Moscow launched a study of the synthesis of amino acids, pro-

Judith, Peter, and Derek Stone enjoy a visit to Great Keppel Island (Queensland) in 1982.

teins, and other substances of nutritional value. By the time of my second visit to his Institute in 1978 he was able to offer visitors black caviar synthesized by his team. I found the taste indistinguishable from the real thing, but I am not a connoisseur of caviar.

If one has made scientific visits to several different countries, over a period of many years, it is certain that incidents will have occurred that require at the time of their occurrence an unusual degree of sang-froid. All speakers encounter the occasional breakage of lecture slides in projectors that malfunction, or the loss of electrical power when giving a talk in a Third World country. Such events are not unusual, but occasionally one is blessed with experiences that are highly amusing and are worth retelling, particularly if one is left with a slight suspicion in one's mind that one was being set up. Two such incidents are narrated here.

The first took place at a university in South Africa, which shall remain nameless in order to avoid embarrassment, as will

A distinguished group of chemists with Judy Stone and me at the 8th International Conference on Organometallic Chemistry held in Kyoto, Japan, in September 1977. Left to right: Alan Sargeson (Australian National University), Judy Stone, Martin Bennett (Australian National University), Edward Abel (Exeter University), me, Michael Lappert (Sussex University), and Malcolm Green (Oxford University). The photograph was taken by Professor Sei Otsuka when we were his guests for a luncheon.

the name of the professor since retired who was my host. Having handed my slides over to the projectionist earlier, my host led me to the lecture theater in his department a few minutes before the talk was due to commence and invited me to take a chair in the front row, where we sat as the room filled with an audience. I was feeling reasonably relaxed, but I did observe that my host was becoming somewhat agitated. Suddenly he leapt to his feet, grabbed my arm, and rapidly dragged me out of the theater, announcing that we were in the wrong room! I often wonder about that audience and how they would have responded to my rising to give a talk on alkeneplatinum complexes. When I eventually arrived in the correct lecture theater I was far from relaxed, and I was left with the slightly uneasy feeling that my host might have deliberately taken me to the wrong theater initially.

India 1984: Nizam for the day at Hyderabad.

The second incident happened in Helsinki. Having arrived in the morning at the departmental office on the third floor of the chemistry building, the chairman began to summarize the projects under way in his own research group. After some minutes the building was rocked by an explosion below our feet, and I naturally expected our conversation to cease immediately while he investigated the cause. However, the summary he was presenting of his work continued unabated. I was impressed by his total lack of concern at what I imagined must have been a serious accident in his laboratory, but I attributed his relaxed attitude to the stalwartness of the Finnish character. After some 20 minutes I regained my composure, whereupon my chair was rocked by a second explosion seemingly of even greater magnitude than the first. Again there was no interruption in the flow of words from my host. Not being able

Relaxing in Hawaii in 1989 with our friend Martin Bennett (center) of the Australian National University. Martin is one of several well-known organometallic chemists who spent study leaves with me in Bristol.

to contain my anxiety further, I asked if we should not investigate the cause of the explosions. With a charming smile my host informed me that a tunnel for the Helsinki subway system was under construction beneath the chemistry building. The dynamiting continued throughout the day. It would have been nice to have been informed of this construction project on my arrival.

Metal Cluster Chemistry

Heteronuclear Metal–Metal Bonds

The domain of metal cluster chemistry now embraces countless polynuclear metal complexes, since the transition elements show a remarkable propensity to form complex compounds with metal–metal bonds. This large field[160,161] is wide in scope and includes many inorganic species in which the metal centers do not carry organic ligands and are therefore not organometallic complexes. However, numerous organometallic compounds contain metal–metal bonds, and several have been mentioned previously in this book. One now-large subsection of this field concerns those metal complexes having bonds between dissimilar transition elements.[125] The first species of this type to be described[162] appears to have been the anion $[FeCo_3(CO)_{12}]^-$, while at about the same time the dimetal compounds $[MoW(CO)_6(\eta^5\text{-}C_5H_5)_2]$,[163] $[FeNi(CO)_3(\eta^5\text{-}C_5H_5)_2]$,[164] $[MoFe(CO)_5(\eta^5\text{-}C_5H_5)_2]$,[165] and $[FeCo(CO)_6(\eta^5\text{-}C_5H_5)]$[166] were reported.[167] Although these early complexes would now be regarded as unremarkable, their synthesis laid the foundations for a large area of study. It would not have been imagined in the early 1960s that within 30 years more than 2000 complex compounds with heteronuclear metal–metal bonds would be described. Moreover, many important reactions are catalyzed heterogeneously by bimetallic systems, for example, hydrocarbon skeletal rearrangements, hydrogenation and isomerization of carbon–

carbon multiple bonds, hydrogenation of CO, hydroformylation, and hydrodesulfurization.

Since our early work in 1960,[165] I had retained an interest in this class of compound, and the availability of the alkene–Pt^0 complexes provided a stimulus for further studies. The discovery, mentioned earlier,[141,168] that the platinum compounds would undergo oxidative insertions into the cage structures of both *closo-* and *nido-*carboranes suggested that these same reagents would also combine with polynuclear metal complexes that are formally electron deficient. The triosmium compound $[Os_3(\mu\text{-}H)_2(CO)_{10}]$ (**72**), discovered by Johnson, Lewis,* and Kilty,[169] provided a good model to test this idea because it contained two three-center, two-electron Os–H $\longrightarrow$ Os bonds, similar to the B–H $\longrightarrow$ B bonds present in diborane and in the *nido-*carboranes we had used in earlier work. Compound **72** has 46

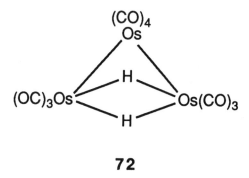

72

cluster valence electrons rather than the 48 of electronically saturated $[Os_3(CO)_{12}]$. The "magic number" 48 in the latter allows each metal–metal connectivity to be assigned an electron pair, and each connection is therefore a bond, whereas with 46 electrons this is not possible in the bridge system of $[Os_3(\mu\text{-}H)_2(CO)_{10}]$. The project was enthusiastically taken up in 1977 by my very productive graduate student Louis Farrugia, now actively researching at Glasgow University, who found that

---

*　Most readers are surely aware of Jack Lewis's many important contributions to inorganic chemistry, initially in association with Ronald Nyholm at University College, London; his distinguished researches have continued at Cambridge for many years. Both they and Joseph Chatt were leaders in the renaissance of inorganic chemistry in Great Britain that began in the 1950s.

compound **72** not only reacted readily with $[Pt(C_2H_4)_2(PR_3)]$, but also afforded nickel–, rhodium–, and gold–triosmium cluster compounds on treatment with the reagents $[Ni(C_2H_4)(PPh_3)_2]$, $[Rh(acac)(C_2H_4)_2]$, and $[AuMe(PPh_3)]$,[170,171] respectively. This method of obtaining "mixed-metal" cluster compounds by adding metal–ligand fragments to electronically unsaturated metal complexes was developed by us[172,173] and has subsequently to my amusement been rediscovered by others, as interest in heteronuclear metal cluster compounds gained momentum at the beginning of the 1980s. The compound $[Os_3(\mu\text{-}H)\{\mu\text{-}Au(PPh_3)\}(CO)_{10}]$, obtained from **72** and $[AuMe\text{-}(PPh_3)]$, was probably the first cluster complex to be reported[170] that demonstrated an electronic relationship between H and $Au(PPh_3)$ groups, serving as a model for numerous other polynuclear metal species later described containing gold–phosphine fragments bridging transition metal atoms in a cluster. Limitations of space allow only a few examples of our own work to be given here to enable the reader to acquire the flavor of this interesting area of research.

The tetranuclear complexes **73** are the products of reactions between $[Pt(C_2H_4)_2(PR_3)]$ and compound **72**. They contain a $Pt(CO)(PR_3)$ group as a consequence of CO migration between metal centers, a common feature in metal cluster chemistry. The compounds **73** are electronically unsaturated, reacting reversibly

73

with CO or H_2 and irreversibly with PR_3 and CH_2 (from CH_2N_2).[174] The products from H_2 and CH_2 retain the metalla-tetrahedrane structure of the precursor 73, but those from CO and PR_3 have a "butterfly" metal core with the platinum atom at a wingtip site. The opening and closing of cluster structures, as a result of adding small molecules, is believed to be relevant to certain catalytic reactions, such as the water gas shift reaction.[175]

In another study[176] we observed that the diruthenium compounds $[Ru_2(\mu\text{-}CH_2)(\mu\text{-}CO)(CO)(L)(\eta^5\text{-}C_5H_5)_2]$ (L is CO or NCMe) react under mild conditions with the platinum reagents $[Pt(C_2H_4)_2(PR_3)]$ (R is cyclo-C_6H_{11} or Pr^i) to afford ruthenium-platinum cluster compounds including the tetranuclear metal species $[Ru_2Pt_2(\mu\text{-}H)(\mu_4\text{-}CH)(\mu\text{-}CO)(CO)_2(PR_3)_2(\eta^5\text{-}C_5H_5)_2]$ and $[Ru_2Pt_2(\mu\text{-}H)_2(\mu_4\text{-}C)(\mu\text{-}CO)_2(PR_3)_2(\eta^5\text{-}C_5H_5)_2]$. These reactions serve to demonstrate a stepwise conversion of CH_2 groups into CH and C ligands at heteropolynuclear metal centers, processes that are the reverse of those invoked to account for products in Fischer–Tropsch chemistry. Moreover, the species that contain μ_4-CH groups have structures in which the C–H bond points away from a ring of four metal atoms (Figure 3). There is thus

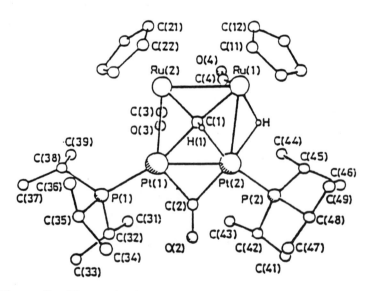

Figure 3. The molecular structure of $[Ru_2Pt_2(\mu\text{-}H)(\mu_4\text{-}CH)(\mu\text{-}CO)(CO)_2(PPr^i)_2(\eta^5\text{-}C_5H_5)_2]$. (Reproduced with permission from reference 176. Copyright 1990.)

an analogy with a C_1 fragment located on the fourfold symmetry axis of the (100) surface plane of a body-centered cubic metal. The structure is to our knowledge unique at the present time, but it is only one of many discovered by organometallic chemists in recent years that serve as models for surface organic chemistry.

By 1979, for reasons given in the next section, our strategies for the synthesis of compounds with heteronuclear metal–metal bonds had become strongly influenced by the "isolobal model" developed by Roald Hoffmann,[68] to which I have briefly referred earlier. In the early 1960s many researchers began to appreciate similarities in behavior between different metal–ligand fragments, and also between particular metal–ligand groups and organic radicals or carbenes. It remained for Hoffmann[68] to sharpen these ideas and show that these "isolobal" relationships arose because the various groups had similar frontier orbitals with the same number of valence electrons, for example,

$$Mn(CO)_5 \longleftrightarrow CH_3 \longleftrightarrow Fe(CO)_2(\eta^5\text{-}C_5H_5)$$

$$Fe(CO)_4 \longleftrightarrow CH_2 \longleftrightarrow Co(CO)(\eta^5\text{-}C_5H_5)$$

$$Co(CO)_3 \longleftrightarrow CH \longleftrightarrow Ni(\eta^5\text{-}C_5H_5)$$

Roald Hoffmann, whom I had first come to know at Harvard, was at this stage of our work in regular contact, generously sending us preprints of his articles, and we were particularly intrigued to receive a paper[177] pointing out the similarity of the frontier orbitals of $[Rh_2(\mu\text{-}CO)_2(\eta^5\text{-}C_5H_5)_2]$ with those of ethylene. This rhodium compound did not exist, so its chemistry could not be explored, but the related compound $[Rh_2(\mu\text{-}CO)_2(\eta^5\text{-}C_5Me_5)_2]$ (**74**) had been prepared by Peter Maitlis and one of his students.[178] The preprint from the Cornell group reached Bristol in 1979, about the same time that Geoffrey Pain joined me as a postdoctoral assistant from Monash University, Australia. He set about showing that compound **74** can, like C_2H_4, form "complexes" with metal–ligand fragments. Thus the "pseudo-alkene" **74**, when treated with $[Mo(CO)_5(thf)]$, $[Fe_2(CO)_9]$, and $[Pt(cod)_2]$, yielded the complexes **75**.[179] Since

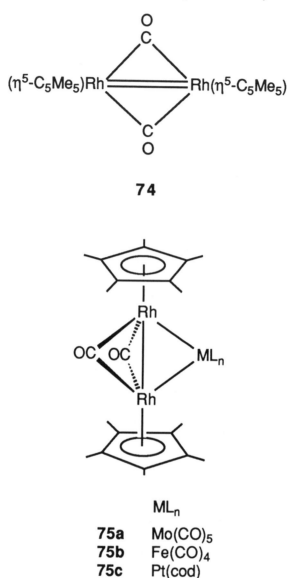

74

ML_n

75a	$Mo(CO)_5$
75b	$Fe(CO)_4$
75c	$Pt(cod)$

the ML_n fragments in the latter are isolobal with CH_2, the products can be regarded as trimetallacyclopropanes. It is therefore
not surprising that **74** also reacts with CH_2N_2 to afford compound **76**. Strikingly, compound **74** displaced the ethylene
groups from $[Pt(C_2H_4)_3]$ to give the novel pentanuclear metal
complex **77**.[180] Crystals of the latter can be monoclinic or

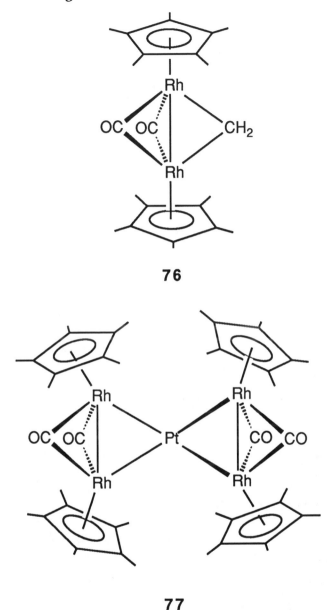

76

77

orthorhombic, but in both structures the rhodium atoms are arranged around the platinum in a pseudo-tetrahedral manner (Figure 4). The four carbonyl ligands bridging the Rh–Rh bonds interact weakly with the platinum atom. The organic molecule equivalent to **77** is spiropentane.

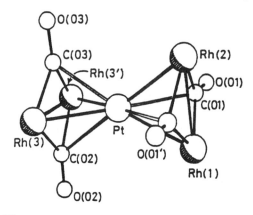

Figure 4. The core atoms in the orthorhombic form of [PtRh$_4$(μ-CO)$_4$(η^5-C$_5$Me$_5$)$_4$] (77). The CO ligands triply bridge the PtRh$_2$ triangles asymmetrically.

Roald Hoffmann with me in Vienna, 1985, attending the XIIth International Conference on Organometallic Chemistry.

Independent of our studies with the dirhodium compound **74**, which as mentioned were initiated in 1979,[181] Larry Dahl and his co-workers[182] had shown that $[Co_2(\mu\text{-}CO)_2(\eta^5\text{-}C_5Me_5)_2]$, the cobalt analog of **74**, would add photochemically generated metal–ligand fragments isolobal with CH_2. Thus a natural development was our syntheses of the mixed-metal "inorganic alkene" $[CoRh(\mu\text{-}CO)_2(\eta^5\text{-}C_5Me_5)_2]$, which on treatment with $[Mo(CO)_5(thf)]$ gave the "trimetallacyclopropane" **78**.

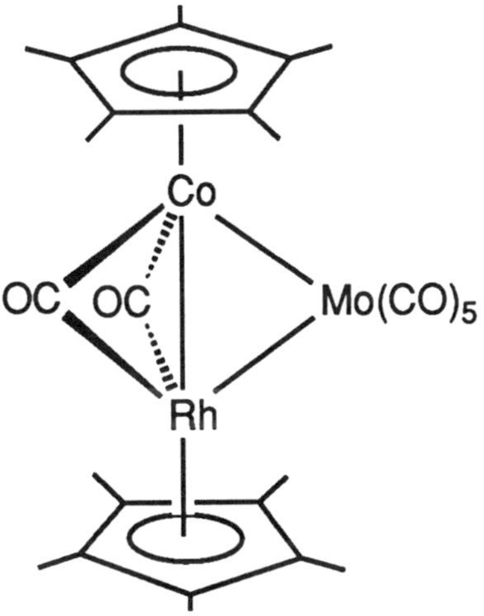

78

The latter was prepared[183] to demonstrate the existence, via variable-temperature NMR spectroscopy, of dynamic behavior akin to that commonly found in metal–alkene complexes; the alkene fragment $CoRh(\mu\text{-}CO)_2(\eta^5\text{-}C_5Me_5)_2$ does indeed rotate about an axis through the Mo and the midpoint of the Co–Rh vector:

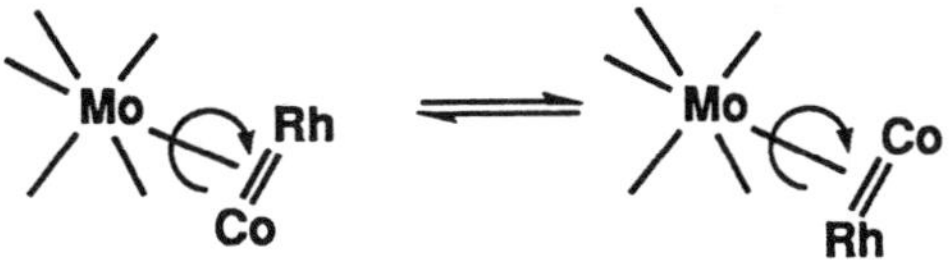

In the low-temperature limiting spectrum, the preferred orientation of the Co–Rh vector with respect to the four radial carbonyl groups of the square-pyramidal $Mo(CO)_5$ moiety is staggered. The link between the "metal-alkenes" and alkenes is strengthened because the barrier to rotation in **78** ($\sim$52 kJ/mol) is similar to that observed for group 6 metal-(d^6-ML_5) alkene complexes. Todd Marder, a postdoctoral fellow who had studied for his Ph.D. at UCLA under Fred Hawthorne, played a very creative role in our work on mixed-metal complexes in this period. Todd, now with his own productive group at the University of Waterloo, Canada, had more knowledge of the relevant chemical literature than any other of my co-workers or even me! Mention must also be made of Guy Orpen, whom I had recruited for a staff position, and who on several occasions made pivotal contributions to structural aspects of the research, besides developing at Bristol his own independent lines of endeavor. This was a good example of the synergism that existed among the inorganic chemistry staff while I was at Bristol.*

Metal–Metal Bonds Bridged by Alkylidene or Alkylidyne Groups

Fischer and his co-workers[184] pioneered the development of the chemistry of metal–carbene and metal–carbyne complexes with the synthesis of species such as [W{=C(OMe)Ph}(CO)$_5$] and [W($\equiv$CPh)(Br)(CO)$_4$]. The carbene- (or alkylidene-) metal complexes were reported in 1964,[185] and the carbyne- (or alkylidyne-) metal complexes later in 1973.[186] Much earlier the Russian chemist Chugaev had prepared platinum complexes with alkylidene ligands, but the true nature of these species had not been recognized.[27] It was Fischer's work that opened up the field, but considerable further stimulus was provided by Richard Schrock's[187] discovery of complexes containing methylene and methylidyne ligands, for example, [Ta(=CH$_2$)(Me)(η^5-C$_5$H$_5$)$_2$] and [W($\equiv$CH)(Cl)(PMe$_3$)$_4$]. Many hundreds of complexes are now known wherein alkylidene or alkylidyne groups are bonded to a single metal center, and the field has been exten-

---

* My younger colleagues M. Green, S. A. R. Knox, A. G. Orpen, and P. L. Timms all received the Corday-Morgan Medal and Prize of the Royal Society of Chemistry during my tenure as head of inorganic chemistry at Bristol.

sively reviewed.[188–190] Alkylidene–metal compounds have been implicated in alkene metathesis and employed in a variety of organic syntheses, while alkylidyne–metal compounds are involved in alkyne metathesis and polymerization and, as described later, may be used in the systematic preparation of metal cluster compounds.[191]

In contrast with the large number of mononuclear alkylidene– and alkylidyne–metal complexes characterized by 1977, relatively few di- or polynuclear metal compounds had been reported with their metal–metal bonds bridged by alkylidene or alkylidyne ligands. The species of this type that had been prepared had been obtained serendipitously. I have already referred to the discovery of compound **71a**. Many other examples of accidental discovery could be given, but three are relevant to the story to be unfolded in this section. The first of these occurred during pioneering studies on reactions of $[Co_2(CO)_8]$ with alkynes, when Sternberg and Wender and their co-workers[192] isolated the tricobalt compound **79**. The cobalt triangle is capped by an ethylidyne fragment, as was subsequently conclusively shown by an X-ray diffraction study made by Sutton and Dahl.[193] Many tricobalt compounds similar to **79**, with different μ_3-CR groups, have been prepared, and their extensive organic chemistry has been studied, especially by Dietmar Seyferth and his co-workers.[194] In a second important discovery, Fischer and Beck[195] isolated the trinickel compound **80** by treating $[Mo\{=C(OMe)Ph\}(CO)(NO)(\eta^5\text{-}C_5H_5)]$ with $[Ni(CO)_4]$. Apparently the fact that the transfer of the carbene group from Mo to Ni might have occurred via an intermediate $Mo\{\mu$-

79

80

C(OMe)Ph}Ni species escaped notice. Third, while attempting to prepare the methylene complex $[Mn(=CH_2)(CO)_2(\eta^5\text{-}C_5H_5)]$ by treating $[Mn(CO)_2(thf)(\eta^5\text{-}C_5H_5)]$ with CH_2N_2, Wolfgang Herrmann and co-workers[196] isolated complex **81**, the first dimetal compound with a μ-methylene bridge. This was a very important discovery, but the idea that **81** might have been formed by addition of a "carbene-like" fragment $Mn(CO)_2(\eta^5\text{-}C_5H_5)$ to the $Mn=CH_2$ group of an intermediate mononuclear manganese species $[Mn(=CH_2)(CO)_2(\eta^5\text{-}C_5H_5)]$ seems to have been overlooked at the time. Anyhow, I was not reading Wolfgang Herrmann's papers with the attention they deserved.

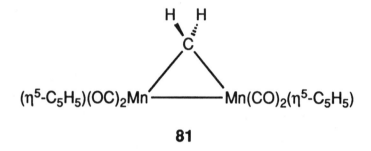

81

By the fall of 1977, Louis Farrugia had successfully added[170] nucleophilic metal–ligand fragments to compound **72**, and consequently the possibility of a strategy for cluster synthesis based on the process

was very much in mind long before Geoff Pain worked with compound **74**, as described in the previous section. Moreover, even earlier[197] Antonio Laguna obtained compounds **71a** and **71b** as part of detailed studies on the chemistry of $[Pt(cod)_2]$. As mentioned previously, formation of **71a** and **71b** had been rationalized in terms of the intermediacy of dipolar $M^+\text{--}^-C(CF_3)_2$ groups that combined with M(cod) fragments to yield the final products. Because the ylid form $M^+\text{--}^-C(CF_3)_2$ could equally well be formulated as $M=C(CF_3)_2$ and also influenced by Farrugia's project and the plethora of platinum–alkene and platinum–alkyne complexes we had made, I was led to think in terms of a general synthetic procedure based on the methodology depicted in Chart II.[198]

It followed immediately that suitable mononuclear metal–alkylidene or metal–alkylidyne compounds might react with Pt^0 complexes to generate species having heteronuclear metal–metal bonds with bridging carbene or carbyne groups. This idea was quickly supported by experiment. The initial results were timely because they came in a period of rapidly growing interest in the reactivity of C_1 fragments bridging metal–metal bonds. As mentioned earlier, when our μ_4-CH

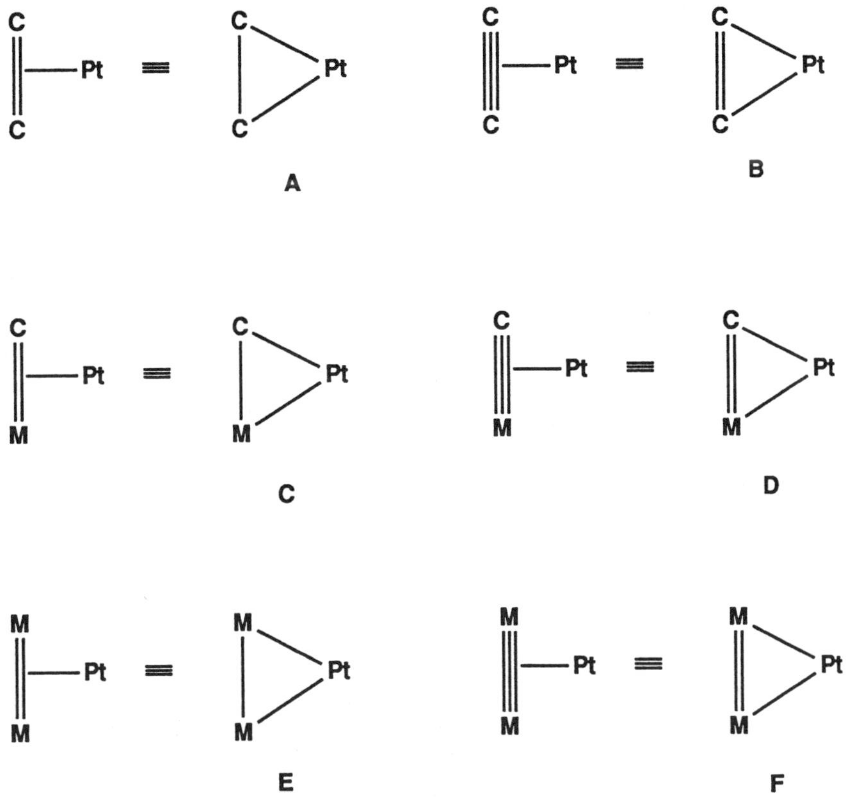

Chart II. A simplistic but conceptually useful view of the "complexation" of Pt-ligand fragments with double and triple bonds (ligands on metal centers have been omitted for clarity). (Reproduced with permission from reference 198. Copyright 1979.)

dirutheniumdiplatinum clusters were discussed, interest in ligated C_1 fragments was stimulated by the experimental evidence that surface-bound CH, CH_2, and CH_3 groups are involved in many heterogeneously catalyzed reactions, especially Fischer–Tropsch syntheses.[199–202] It was thought likely that the reactivity patterns shown by these C_1 groups when bridging metal–metal bonds in discrete molecular complexes might mimic their behavior on metal surfaces. However, great care has to be taken in drawing such a mechanistic parallel.[201,203] Irrespective of the validity of comparisons made between the behavior of organic fragments as ligands in discrete complexes and their reactivity on metal surfaces, the concept of an analogy has in part been responsible for the discovery of much new chemistry occurring at di- or trimetal centers. In this context the reader is referred to exciting work emanating from the groups of Bergman,[204] Casey,[205] Chisholm,[206] Herrmann,[201] Knox,[207] Norton,[208] Puddephatt,[209] and Shapley.[210] It must be emphasized, however, that the research of these groups has largely focused on homonuclear di- or trimetal compounds, whereas that described here relates to heteropolynuclear metal complexes.

Fortunately, at about the time the significance of the methodology for synthesis based on Chart II became apparent to me, Terence Ashworth, a highly competent chemist from South Africa, joined my group as a postdoctoral fellow. His previous Ph.D. studies with Eric Singleton at the National Chemical Research Laboratory in Pretoria had given him good experience in the synthesis of organometallic compounds, and he enthusiastically accepted my ideas. Moreover, he made them work. He immediately set about treating Fischer's complex $[W\{=C(OMe)Ph\}(CO)_5]$ with $Pt(PR_3)$ and $Pt(PR_3)_2$ fragments derived from the reagents $[Pt(C_2H_4)_2(PR_3)]$ and $[Pt(C_2H_4)(PR_3)_2]$ or $[Pt(PR_3)_2]$ (R is a bulky group), as described earlier. In initial experiments the phosphines used were PEt_3 and PBu^t_2Me. Both were unfortunate choices because platinum fragments containing the former ligand produced insoluble products, making NMR studies difficult, while the compound $[Pt(PBu^t_2Me)_2]$ afforded the trimetal complex 82, a product resulting from loss of a bulky PBu^t_2Me group from platinum and transfer of the carbene ligand from tungsten to the Pt–Pt bond. Formation of complex 82 was of course encouraging since it was reminiscent of Fischer's synthesis of the trinickel compound 80, referred to earlier.

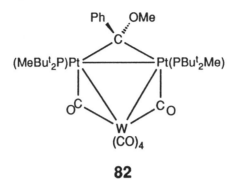

82

The breakthrough came in the spring of 1978 when Ashworth used tertiary phosphines with smaller cone angles than PBu^t_2Me and prepared the complexes $[Pt(C_2H_4)(PR_3)_2]$ (where PR_3 is PMe_3 or PMe_2Ph) from $[Pt(C_2H_4)_3]$. It was then shown that treatment of $[W\{=C(OMe)Ph\}(CO)_5]$ with $[Pt(C_2H_4)(PMe_3)_2]$ afforded compound **83a**, the first unequivocally characterized dimetal compound of the type we were seeking.[211] It was evident that we had hit upon an incredibly versatile method for preparing heteronuclear dimetal compounds in a systematic way, and two other postdoctoral assistants, Madeleine Berry (later Helliwell) from Oxford and Mariano Laguna from Zaragoza, were deployed on the project to ensure rapid progress. Chromium and molybdenum analogs of complex **83a** were quickly made, as were the compounds **84** from $[Mn\{=C(OMe)Ph\}(CO)_2(\eta^5\text{-}C_5H_5)]$.[212] Complexes **83** and **84** can be regarded as "dimetallacyclopropanes" because the metal–ligand fragments are isolobal with CH_2. I shall not discuss further the bridging carbene complexes. We have had little opportunity to exploit

	M		M
83a	W	**84a**	Pt
83b	Mo	**84b**	Pd
83c	Cr	**84c**	Ni

their chemistry, which is undoubtedly interesting, simply because in mid-1978 Terry Ashworth carried out a reaction that gave results so significant that it has influenced most of our subsequent work. Ashworth treated Fischer's well-characterized carbyne compound $[W(\equiv CC_6H_4Me-4)(CO)_2(\eta^5-C_5H_5)]$ with $[Pt-(C_2H_4)(PMe_2Ph)_2]$ and obtained compound **85** in over 80% yield.[198] Within 3 days of being given crystals of **85**, Judith Howard, enthusiastic as usual, had established the structure by X-ray diffraction.

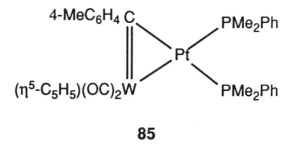

85

In 1957 Chatt and co-workers[213] had been the first to characterize monoplatinum–alkyne complexes and had suggested the formulation depicted in **86**. The relationship between **85** and **86**, based on the electronic equivalence of CR and $W(CO)_2(\eta^5-C_5H_5)$ groups, is immediately apparent. Complex **85** can be regarded as a "dimetallacyclopropene", and this description accounts for much of its chemistry. Judith Howard's X-ray diffraction study established that the μ-C–W distance was that expected for a C=W bond. The $Pt(PMe_2Ph)_2$ fragment is isolobal with CH_2, and it was immediately obvious that numerous

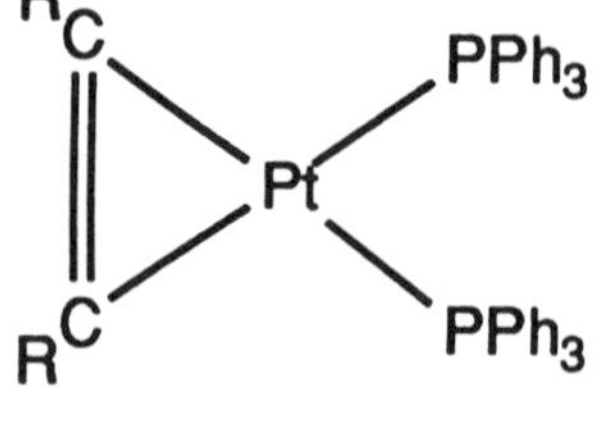

86   R = alkyl or aryl

"complexes" should be formed by treating $[W(\equiv CC_6H_4Me\text{-}4)\text{-}(CO)_2(\eta^5\text{-}C_5H_5)]$ or structurally related species with metal–ligand fragments electronically related to methylene.

Within a relatively short time several excellent co-workers had established the generality of Terry Ashworth's reaction, as indicated in Scheme IV.[191,214] Preparative chemistry is very competitive, and it was necessary to exploit the initial results quickly before others did so. As on earlier occasions when an important breakthrough in research had occurred in my laboratory, I generally encouraged other members of the group to place their own projects on hold, and help in the new experimental work. Such a policy, however, requires action on the part of the supervisor so as to avoid having the person who initiated the area feel that his project is being unfairly taken over by others. One should leave no doubt in the mind of any member of the group that each individual's contribution will not be forgotten. I am glad to say that in Ashworth's case I was in part responsible for his receiving the Raikes Medal from the South African Chemical Institute and a Humboldt Fellowship shortly after he left Bristol. It should be noted that the metals bonded to tungsten or molybdenum in the products (Scheme IV) include elements as diverse as Ti and Cu. Moreover, it is possible to change the ligands in the fragments $M'L_n$, provided the fragments remain isolobal with CH_2. Also, the R group in the reagents $[M(\equiv CR)\text{-}$

$$(\eta^5\text{-}C_5R'_5)(OC)_2M \equiv CR \quad + \quad M'L_n \quad \longrightarrow \quad (\eta^5\text{-}C_5R'_5)(OC)_2M \overset{\displaystyle \overset{R}{\underset{\diagup \diagdown}{C}}}{\text{———}} M'L_n$$

M = W or Mo, R = aryl, alkyl, alkynyl, R' = H or Me; $M'L_n = Cu(\eta^5\text{-}C_5Me_5)$, $Pt(PR_3)_2$, $M'(CO)(\eta^5\text{-}C_5Me_5)$ (M' = Co or Rh), $M'(CO)(\eta^5\text{-}C_9H_7)$ (M' = Rh or Ir), $M'(CO)(acac)$ (M' = Rh or Ir), $Fe(CO)_n$ (n = 3 or 4), $Mn(CO)_2(\eta^5\text{-}C_5H_4Me)$, $Re(CO)_2(\eta^5\text{-}C_5H_5)$, $Cr(CO)_2(\eta^6\text{-}C_6Me_6)$, $Cr(CO)(NO)(\eta^5\text{-}C_5H_5)$, $M'(\mu\text{-}\eta^2\text{-}CO)(\eta^5\text{-}C_5H_5)_2$ (M' = Ti or Zr)

Scheme IV. The synthesis of dimetal compounds with W or Mo bonded to other transition metals and post-transition elements. The metal-ligand fragments are isolobal with CH_2, and the $M\equiv C$ bond functions formally as a two-electron donor like a $C\equiv C$ group. However, in some dimetal species it has been established that a $M\equiv C$ bond may act as a three- or four-electron donor, as can $C\equiv C$ bonds in certain alkynemetal complexes (251).

$(CO)_2(\eta^5\text{-}C_5H_5)]$ may be varied from $C_6H_4Me\text{-}4$ to Me, C_6H_4-$OMe\text{-}2$, $C_6H_3Me_2\text{-}2,6$, Ph, or even $C{\equiv}CR$, $SiPh_3$, and so forth, and the ligand $\eta^5\text{-}C_5H_5$ may be replaced by $\eta^5\text{-}C_5Me_5$.

The alkylidyne–cyclopentadienyltungsten or -molybdenum reagents **A** and **B** of Chart III, together with the cyclopentadienyl-free alkylidyne–metal species **C** to **F**, allowed the preparation of a host of dimetal compounds by the most systematic method yet devised. Moreover, from the standpoint of organic chemistry, the various products obtained provide an excellent opportunity to study the reactivity of alkylidyne groups bridging dissimilar metal atoms. The advantage of such species over the homonuclear dimetal compounds lies in having the C_1 fragments associated with metal centers possessing different ligating properties. This feature can enhance the reactivity of the unsaturated three-membered rings and also allows extensive "fine tuning" by changing the ligands on one or other metal center.

Over a period of a few years I was able to attract for this work several excellent postdoctoral assistants, including Gabino Carriedo and Miguel Ruiz from Professor Riera's group at Oviedo, Diane and Greg Lewis from Al Cotton's group at Texas A & M, and Tony Hill from Max Herberhold's group at Bayreuth. Tony had started his research with Warren Roper in Auckland, New Zealand, before moving to Germany, so he was a very experienced worker. Also, Esther Delgado joined us from Madrid, and Peter Byers from Tasmania. In attracting top-class young chemists it appears that nothing succeeds like a good track record, and I do not recall ever having to advertise a postdoctoral vacancy. All contributed significantly to the research program, and with the exception of the Lewises, who subsequently joined Dow Chemical, all have embarked on promising academic careers.

It should not be construed from this comment that it was ever my aim to force co-workers into an academic career. Many of my past associates, both Ph.D. students and postdoctoral fellows, have been very successful in industrial positions or in government employment. One hears far too much from some people about young researchers being brainwashed into following in the footsteps of their mentors with their only aim being publication in the primary literature. My co-workers have had sufficient individuality to make up their own minds as to

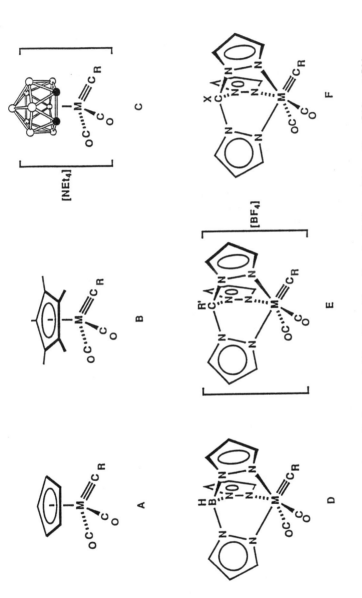

M = Mo or W, R = alkyl or aryl, R' = H or Me, X = AuC$_6$F$_5$ or BF$_3$, ⊚ = CMe or CH

Chart III. A family of complexes with M≡CR groups that readily add metal–ligand fragments to form products with metal–metal bonds and bridging alkylidyne groups.

Professors Max Herberhold (Bayreuth), at right, and Heinz Nöth (München) at Wilbad-Kreuth, Bavaria, September 1992. The occasion was a conference commemorating 20 years since the discovery in Fischer's laboratory of metal complexes with carbon-to-metal triple bonds. Max Herberhold and Heinz Nöth have been at the forefront of research in inorganic chemistry in Germany for many years.

whether they wish to pursue a career in industry or in academia. Having first made a decision as to which avenue to follow, most of my young colleagues usually sought help as to how to go about accomplishing their aims. Of course many Ph.D. students are uncertain about which route to follow after completing their research, and in these circumstances it is natural to advise them to consider a postdoctoral position, so as to keep their options open.

Scheme V summarizes a few reactions of just one dimetal compound, studied by Esther Garcia,[215] another postdoctoral fellow from Oviedo. These reactions are included to illustrate C–C, C–O, and C–S bond-forming processes involving the alkylidyne group. In some cases these processes are simple in the sense of affording a single product, for example, that with

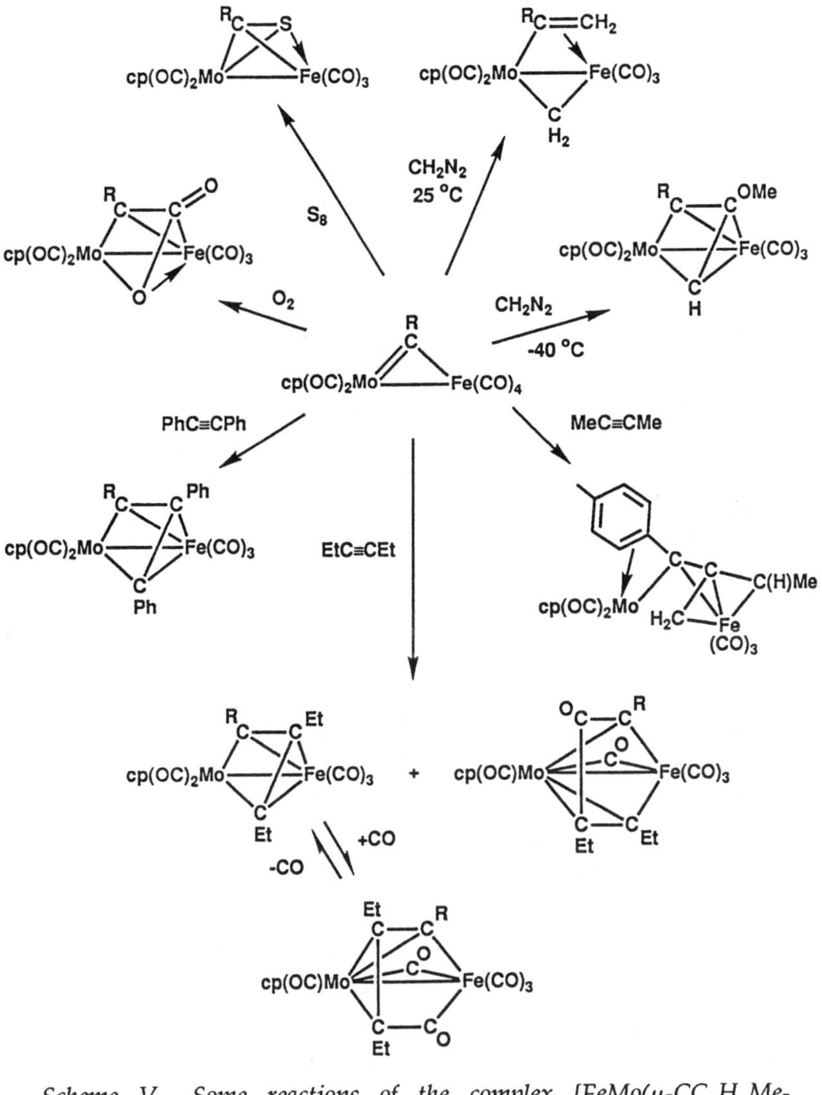

Scheme V. *Some reactions of the complex [FeMo(μ-CC$_6$H$_4$Me-4)(CO)$_6$(η^5-C$_5$H$_5$)], which serve to illustrate C–C, C–O, and C–S bond-forming processes at a heteronuclear dimetal center. R is C$_6$H$_4$Me-4, and cp is η^5-C$_5$H$_5$.*

oxygen. In other instances the temperature at which the reaction is carried out determines the nature of the product, for example, that with CH$_2$N$_2$. Reactions at low temperatures evidently increase the lifetime of certain of the intermediates

The 1985 group of postdoctoral assistants from Spain. Left to right, back row: Drs. Miguel Ruiz, Fernando Mayor-Real, Esther Delgado, Daniel Miguel, and Gabriel Garcia. Front row: Dr. Dolores Bermudez, me, and Dr. Esther Garcia.

involved, thus allowing different pathways to be followed. In other reactions, for example with 3-hexyne, mixtures of products are obtained irrespective of temperature. Some of the reactions depicted in Scheme V involve molecular rearrangements at a metal center, including hydrogen migrations. Space limitations do not allow discussion of possible pathways to the various products, many of which have been established by ^{13}C or ^{2}H labeling experiments, but one example may be given. By ^{13}CO labeling of the carbonyl groups in $[FeMo(\mu\text{-}CC_6H_4Me\text{-}4)(CO)_6(\eta^5\text{-}C_5H_5)]$ it was possible to demonstrate that the CO group of the COMe fragment in the product obtained from CH_2N_2 at -40 °C, namely, $[FeMo\{\mu\text{-}C(C_6H_4Me\text{-}4)C(OMe)C(H)\}(CO)_5(\eta^5\text{-}C_5H_5)]$ was derived from a carbonyl ligand in the precursor. Moreover, by using CD_2N_2 in place of CH_2N_2, it could be demonstrated by ^{2}H NMR that the μ-CH fragment and the Me of the OMe group came from diazomethane.

The reactions of $[FeMo(\mu\text{-}CC_6H_4Me\text{-}4)(CO)_6(\eta^5\text{-}C_5H_5)]$ depicted in Scheme V are not possible with the tungsten analog because the analog is unstable, disproportionating to yield the cluster compounds **87** and **88**. However, although [FeW(μ-

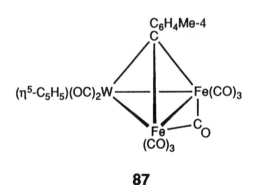

87

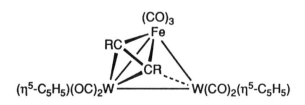

88 R = C_6H_4Me-4

$CC_6H_4Me-4)(CO)_6(\eta^5-C_5H_5)]$ cannot be used in synthesis, because it so readily transforms into trimetal species, the related compound $[FeW(\mu-CC_6H_4Me-4)(CO)_6(\eta^5-C_5Me_5)]$ is stable and does react with CH_2N_2, alkynes, and other reagents to give a variety of products.[216] This illustrates the important feature of "fine-tuning" that is possible with these complexes. In the Fe–W system, replacement of $\eta^5-C_5H_5$ by $\eta^5-C_5Me_5$ produces a stable but very reactive dimetal compound, but the products obtained from $[FeW(\mu-CC_6H_4Me-4)(CO)_6(\eta^5-C_5Me_5)]$ and the various reactants are not in every instance structurally similar to those obtained from $[FeMo(\mu-CC_6H_4Me-4)(CO)_6(\eta^5-C_5H_5)]$. These observations show clearly the potential for further research. There are many heteronuclear dimetal combinations available for study with, as mentioned, the additional flexibility of modifying the reactivity at either metal center by varying the ligands. Some of the reactions that occur at these heteronuclear dimetal centers have been reviewed by my colleagues John Jeffery and Michael Went,[217] who with the others mentioned contributed much valuable experimental work and ideas to our research.

Metal Cluster Compounds with Bridging Alkylidyne Ligands

The analogy between the ligating properties of the compounds $[M(\equiv CR)(CO)_2(\eta^5-C_5H_5)]$ (M is Mo or W; R is aryl, alkynyl, or alkyl) and alkynes, elaborated in Scheme IV and described earlier, pointed the way to the rational synthesis of trimetal compounds with capping alkylidyne groups in an exciting and entirely new way. Thus it was known for many years that $[Co_2(CO)_8]$ and certain other dimetal carbonyl complexes react with alkynes to yield products in which the alkynes transversely bridge the metal–metal bonds, for example, compound **89**. Similarly, the more recently discovered $[Mo_2(CO)_4(\eta^5-C_5H_5)_2]$ affords products such as **90** with alkynes. The existence of a family of dicobalt–alkyne compounds like **89** led us to investigate the reaction between $[W(\equiv CC_6H_4Me-4)(CO)_2(\eta^5-C_5H_5)]$ and $[Co_2(CO)_8]$. Complex **91** was thereby isolated in quantitative yield.[218] The relationship between the structures of **89** and **91** is apparent, as is that between these species and **79**. Similarly, $[W(\equiv CC_6H_4Me-4)(CO)_2(\eta^5-C_5H_5)]$ and $[Mo_2(CO)_4(\eta^5-C_5H_5)_2]$ in toluene at ~100 °C afford **92**,[219] which is structurally akin to compound **90**.

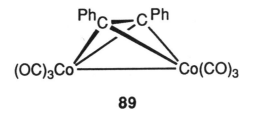

89

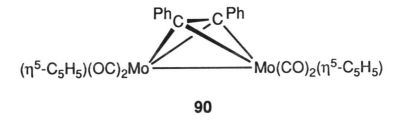

90

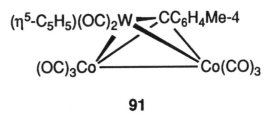

91

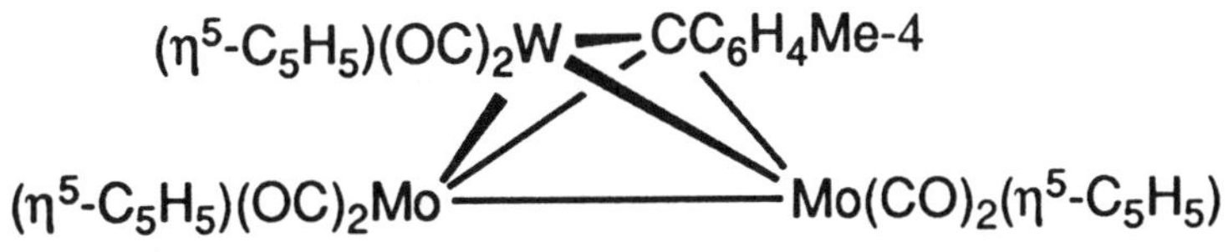

92

An even more versatile approach to trimetal clusters with capping alkylidyne groups involves treating dimetal compounds of the kind shown in Scheme IV with metal–ligand fragments. The possibilities seem almost limitless,[191] but space allows only two examples to be given. Treatment of [RhW(μ-CC$_6$H$_4$Me-4)-(CO)$_3$(η^5-C$_5$H$_5$)(η^5-C$_9$H$_7$)] with [Fe$_2$(CO)$_9$] gives compound **93**, a prochiral cluster containing metals from the first, second, and third transition element series.[220] Addition of Cu^I chloride to a tetrahydrofuran suspension of LiC$_5$Me$_5$ at −78 °C affords a highly reactive solution presumed to contain [Cu(thf)(η^5-C$_5$Me$_5$)]. The latter is a source of the "Cu(η^5-C$_5$Me$_5$)" fragment, isolobal with CH$_2$, and with [PtW(μ-CC$_6$H$_4$Me-4)(CO)$_2$(PMe$_3$)$_2$-(η^5-C$_5$H$_5$)] the copper reagent yields the trimetal compound **94**.[221] Whereas **93** is a "trimetallatetrahedrane" with a closo structure, compound **94** adopts a so-called "butterfly" structure with the Cu and Pt atoms at the "wingtip" positions. This structural difference is a natural consequence of the differing electronic requirements of the metals in the two clusters.

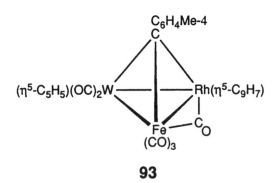

93

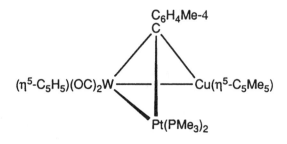

94

The synthesis of numerous compounds with μ_3-CR ligands, such as 91–94, prepared in high yield in a systematic manner, provides chemists with an excellent opportunity to study the reactivity of alkylidyne groups bridging three metal centers. Moreover, association of the μ_3-CR (R is alkyl or aryl) fragments with metals having different bonding requirements leads to variations in the reactivity patterns observed. As with the dimetal compounds shown in Scheme IV, the possibilities for further research seem boundless.

Perhaps the most fascinating development emerged from the isolation of compounds 95a–95c. The paradigm molecule 95a was prepared by Terry Ashworth by treating $[Pt(C_2H_4)_3]$ with two equivalents of $[W(\equiv CC_6H_4Me-4)(CO)_2(\eta^5-C_5H_5)]$. My graduate student Michael Chetcuti, now on the staff at Notre Dame, made an important contribution with the synthesis of the palladium and nickel species 95b and 95c from $[Ni(cod)_2]$ and $[Pd(norbornene)_3]$, respectively, and later many other molecules of this structural type were obtained.[222] Key X-ray diffraction studies by Judith Howard and Peter Woodward and their students established that these molecules have near-linear W–M–W spines, bridged by carbyne groups and semibridged by CO ligands. The structure of 95a is shown in Figure 5. The two three-membered rings are orthogonal to one another, and overall the structure is strikingly similar to those of the alkyne species 96, prepared earlier (*see* Chart I, A). Later the Mo analogs of 95a and 95c were also prepared. These compounds are unsaturated and may be used as precursors to prepare molecules containing chains or rings of metal atoms in a stepwise manner (Scheme VI).

	M	R
95a	Pt	C_6H_4Me-4
95b	Pd	C_6H_4Me-4
95c	Ni	C_6H_4Me-4

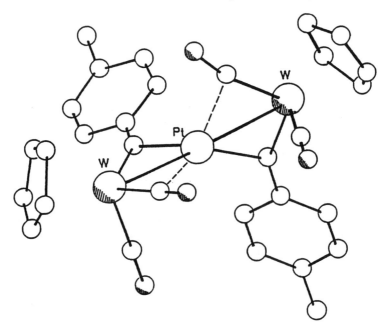

Figure 5.   The molecular structure of the compound [W$_2$Pt(μ-CC$_6$H$_4$Me-4)$_2$(CO)$_4$(η^5-C$_5$H$_5$)$_2$] (**95a**). *The two μ-CPtW rings are orthogonal, and the μ-C—W separations [1.90(1) Å] correspond to that expected for a C=W bond. The dotted lines indicate semibridging CO groups.*

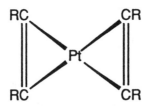

96 R = aryl or alkyl

This may be accomplished in various ways, for example, by adding one or two Pt(cod) fragments (derived from [Pt(cod)$_2$] treated with C$_2$H$_4$) to the trimetal compounds. The cod ligands in the tetra- or pentanuclear metal products may then be displaced by addition of more [M($\equiv$CR)(CO)$_2$(η^5-C$_5$R'$_5$)] (M is Mo or W; R is alkyl or aryl; and R' is H or Me) molecules, so as to

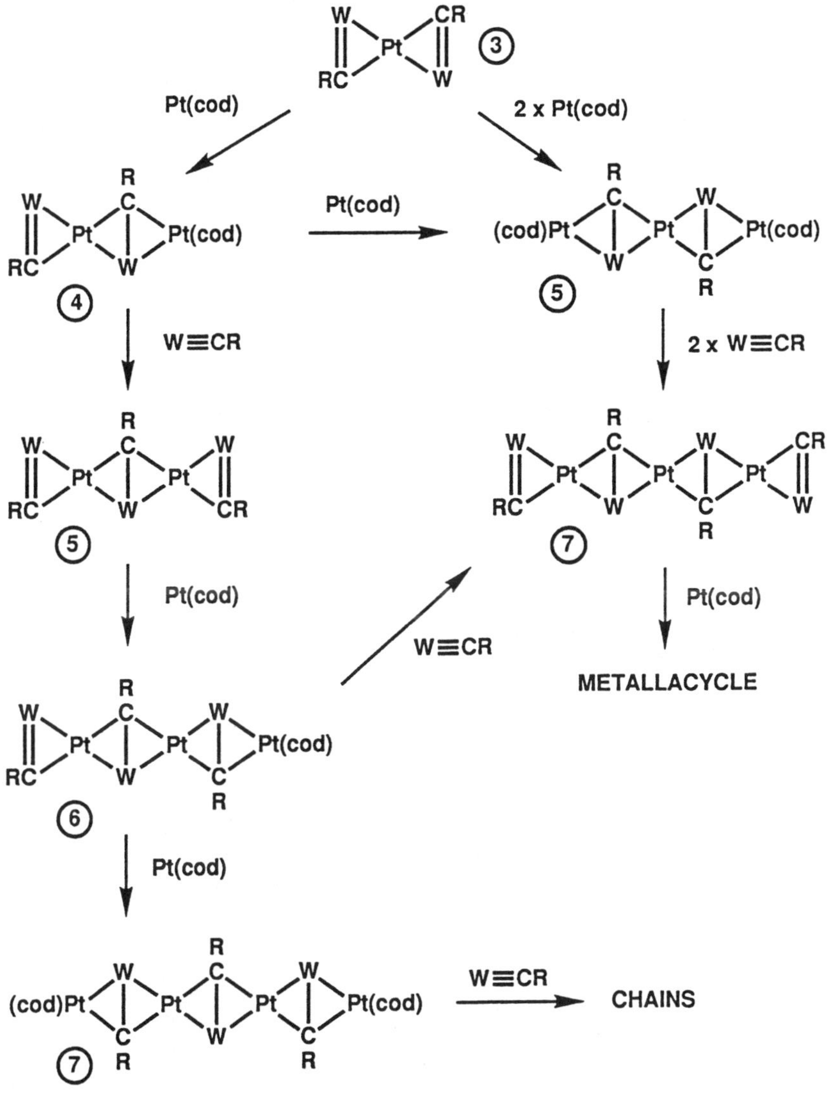

$W\equiv CR$ = [W($\equiv$CR)(CO)$_2$cp]   Pt(cod) = [Pt(cod)$_2$] / C$_2$H$_4$

Scheme VI. Synthetic methodology used to construct metal–chain or metal–ring complexes starting from the species [W$_2$Pt(μ-CR)$_2$(CO)$_4$-(η^5-C$_5$R'$_5$)$_2$] (R is alkyl, aryl, or alkynyl; R' is H or Me). Tungsten may be replaced by molybdenum in these syntheses and platinum by nickel.

increase chain length. The early experiments were carried out by
Iain Moore towards the end of his Ph.D. program, but the work
was significantly extended by Greg Elliott from Warren Roper's
Laboratory in Auckland, New Zealand, and Takaya Mise from
Tokyo, with many later productive studies by Bristol graduate
student Simon Davies.

A brief discussion is given to illustrate the myriad of possibilities for future work, based on the synthetic methodology
developed so far.[223] Treatment of complex **95a** with an excess of

*Two of my co-workers who came from the New Zealand group of the
well-known chemist Warren Roper. The occasion was the 1987 grad-
uation of Greg Elliott, shown on the left, who with careful experimental work was responsible for the discovery of the metal-atom "star clusters". Tony Hill, now a lecturer at Imperial College, had many novel
ideas for research whilst a postdoctoral assistant.*

Warren Roper at Wilbad-Kreuth, Bavaria, September 1992. In 1973, Warren spent a year on study leave with my group at Bristol.

$[Pt(cod)_2]$ in ethylene-saturated thf affords the pentanuclear metal compound **97**. The latter in thf under ethylene (to labilize the cod ligands) yields with an excess of $[W(\equiv CC_6H_4Me\text{-}4)(CO)_2(\eta^5\text{-}C_5H_5)]$ the seven-metal-atom chain complex **98**. Alternatively, a seven-metal-atom chain product may be prepared from a trimetal species. For example, $[W_2Ni(\mu\text{-}CMe)_2(CO)_4(\eta^5\text{-}C_5Me_5)_2]$ with $[Ni(cod)_2]$ gives compound **99**. As expected, the chains can adopt different conformations in solution, with the number of stereoisomers increasing with chain length. Not all possible diastereoisomers exist, but those that do may be detected by NMR spectroscopy (^1H, $^{13}C\text{-}\{^1H\}$, or $^{195}Pt\text{-}\{^1H\}$). Where possible, crystals have been obtained from the solutions, and several X-ray diffraction studies have been carried out by Judith Howard and co-workers. The structure of a six-atom

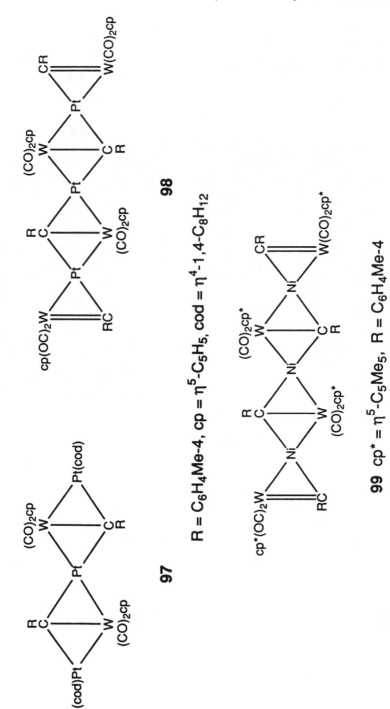

98

$R = C_6H_4Me-4$, $cp = \eta^5-C_5H_5$, $cod = \eta^4-1,4-C_8H_{12}$

97

99 $cp^* = \eta^5-C_5Me_5$, $R = C_6H_4Me-4$

chain cluster is shown in Figure 6. A cod ligand may be displaced from a platinum atom by $[M(\equiv CR)(CO)_2(\eta^5-C_5R'_5)]$ molecules, or, alternatively, chain lengths may be increased by addition of Pt(cod) fragments to molecules containing terminal C=W groups.

Once a chain contains seven metal atoms and has terminal C=W groups, addition of a further metal–ligand fragment usually results in cyclization. For example, the eight-metal-atom complex **100** is formed by treating compound **98** with an ethylene-saturated thf solution of $[Pt(cod)_2]$. Compound **100** has an unsymmetrical structure with one edge-bridging and three triply bridging CC_6H_4Me-4 groups. On heating in solution it rearranges to the symmetrical isomer with four triply bridging CC_6H_4Me-4 ligands, two lying above the puckered eight-metal ring and two below.

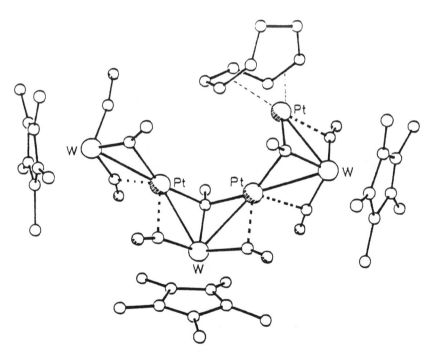

Figure 6. The molecular structure of the six-atom chain compound $[W_3Pt_3(\mu-CMe)(\mu_3-CMe)_2(CO)_6(cod)(\eta^5-C_5Me_5)_3]$, prepared in a stepwise manner by adding one equivalent of $[Pt(cod)_2]$ to $[W_3Pt_2(\mu-CMe)_2(\mu_3-CMe)(CO)_6(\eta^5-C_5Me_5)_3]$.

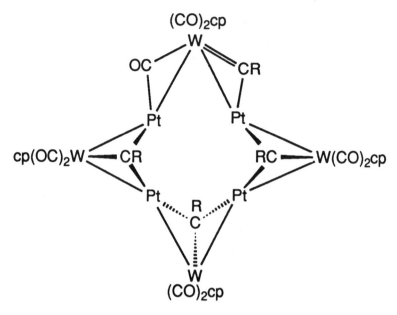

100 R = C_6H_4Me-4

The metallacycles with eight metal atoms form very readily, and if desired they may contain nickel in place of platinum or molybdenum in place of tungsten. Since the metal atom sites conjure up the points of a star, the term "star cluster" has been employed by us to describe these species.[224]

Star clusters containing both nickel and platinum can be prepared in high yield directly from compound **95a**, or its analog containing a phenyl group, by treatment with [Ni(cod)$_2$]. An isomeric mixture is produced consisting of [W$_4$Ni$_2$Pt$_2$(μ_3-CR)$_4$(CO)$_8$(η^5-C$_5$H$_5$)$_4$], having a symmetrical structure with all four CR groups in triply bridging sites alternately above and below the eight-membered metal ring (Figure 7), and two isomers [W$_4$Ni$_2$Pt$_2$(μ-CR)(μ_3-CR)$_3$(CO)$_8$(η^5-C$_5$H$_5$)$_4$] (R is Ph or C$_6$H$_4$Me-4) with unsymmetrical structures. In the latter species three of the CR groups occupy triply bridging sites, while the fourth edge bridges a metal–metal bond. The two isomers differ in that in one the μ-CR group spans a W–Pt bond and in the other a W–Ni bond (Figure 8). On warming mixtures in tetrahydrofuran, both "unsymmetrical" species isomerize to the symmetrical complex with four μ_3-CR groups.[224]

ANGEWANDTE
CHEMIE

Herausgegeben
von der Gesellschaft
Deutscher Chemiker

98/ 2
1986

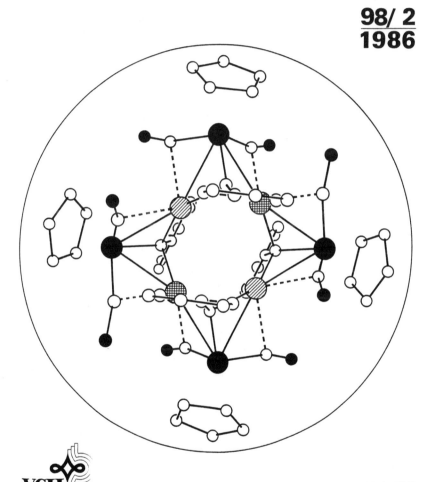

VCH
Verlagsgesellschaft

ANCEAD 98 (2) 115–196 (1986) · ISSN 0044–8249
Vol. 98 - No. 2 - February 1986

Figure 7. The molecular structure of $[W_4Ni_2Pt_2(\mu_3\text{-}CPh)_4(CO)_8(\eta^5\text{-}C_5H_5)_4]$ established by X-ray diffraction. Dark circles are W, crosshatched are Ni, and striped are Pt. Dotted lines indicate semibridging CO ligands. (Reproduced with permission from reference 224a. Copyright 1986.)

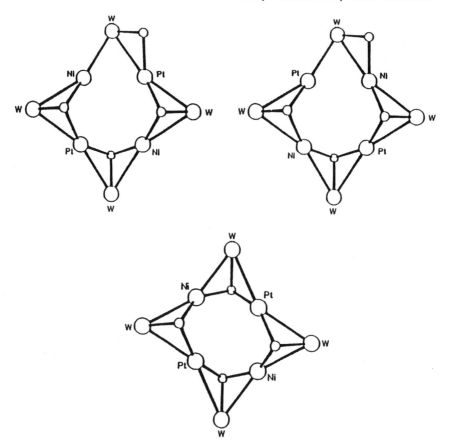

Figure 8. Metal core structures of the "unsymmetrical" [$W_4Ni_2Pt_2$-(μ-CR)(μ_3-CR)$_3$(CO)$_8$(η^5-C$_5$H$_5$)$_4$] and "symmetrical" [$W_4Ni_2Pt_2$(μ_3-CR)$_4$(CO)$_8$(η^5-C$_5$H$_5$)$_4$] isomeric molecules. In one asymmetric "star cluster" the μ-CR ligand bridges a W—Pt bond and in the other a W—Ni bond. In solution the asymmetric species, with one edge-bridging alkylidyne ligand, isomerize into the symmetrical isomer with four triply bridging alkylidyne groups. These groups are sited alternately above and below the plane defined by the eight metal atoms. Experimental procedures may be adapted so as to produce metallacycles with different metal atom sequences, for example, Mo—Pt—W—Ni—W—Pt—Mo—Ni, or W—Pt—Mo—Ni—W—Pt—Mo—Ni.

Work by Simon Davies led to the isolation of products having chains of 9 and 11 metal atoms, for example, compounds **101** and **102**.[225] The structure of an 11-atom metal chain cluster is shown in Figure 9. The starting compounds for these prod-

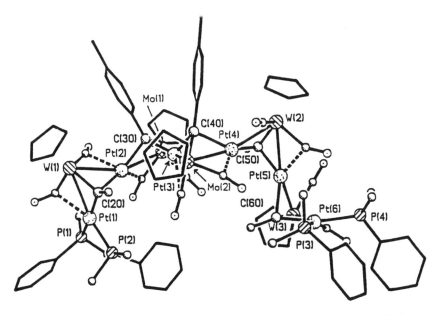

101 R = C_6H_4Me-4

102 R = C_6H_4Me-4

Figure 9. Structure of the 11-metal-atom chain compound $[Mo_2W_3$-$Pt_6(\mu_3\text{-}CMe)_3(\mu_3\text{-}CC_6H_4Me\text{-}4)_2(CO)_{10}(PMe_2Ph)_4(\eta^5\text{-}C_5H_5)_5]$, the cluster of this type with the longest chain of metal atoms yet identified crystallographically. (Reproduced with permission from reference 225b. Copyright 1989.)

ucts are seven-metal-atom chain species having terminal Pt(cod) groups. In principle it should be possible to prepare compounds with even longer metal atom chains and metallacycles having rings with more than eight metal atoms. However, characterization of such species becomes increasingly difficult. With higher molecular weights and decreasing solubility, NMR measurements become more laborious. X-ray diffraction studies therefore become of crucial importance, but it is often very difficult to obtain good-quality crystals. Moreover, there is some evidence that chains with more than eight metal atoms tend to break down, affording the very stable "star clusters". Nevertheless, both the star clusters and the chain structures will provide a multitude of opportunities for further work.

Yes, Minister

At this juncture I shall depart from an account of my research to describe in some detail how, at a time when members of my group were particularly active in the work mentioned in the various sections of this book, I became increasingly occupied by membership of national committees in the United Kingdom concerned with teaching, research funding, and the distribution of resources for chemistry. Under such pressures one has to work even harder in order to maintain "visibility" as a researcher through publications and lectures. There is always the option of refusing the call to serve on committees, and indeed some invitations of this kind must be refused if a proper balance between teaching, research, and administration is to be maintained. However, there is an obligation to join a committee or panel if by so doing one can protect funding and be an advocate for increased resources for colleagues in chemistry throughout the university system. One of several such calls upon my time came in 1987–1988 when I was asked to chair a committee to examine the then-current provision for teaching and research in chemistry in the British universities. The non-British reader will I hope excuse what he or she might regard as the somewhat parochial nature of the following account, which may, however, be instructive as changes occur in policy for scientific research and teaching in the reader's country.

By the mid-1980s the fundamental link between teaching and research in the British universities was being questioned, and the role played by research in institutions of higher education was coming under increasing political scrutiny and criticism. The relevance of much of the scientific research being done was deemed by some Government Ministers, and others who should have known better, not to accord with the perceived economic needs of the country in the short and medium term, and to be not of much use anyway. Scientists came to be seen as a producer cartel, consuming the nation's resources with little thought given to the costs of research in relation to other demands on the public purse. I personally found this attitude to be both ill-judged and surprising since in my opinion scientists are less rather than more vocal than other special interest groups. Politicians seemed incapable of understanding why discoveries in curiosity-driven research do not rapidly contribute positively to a country's gross national product. Deaf ears are turned to arguments that the advancement of knowledge depends on the activities of many, and not solely on the work of a very few major scholars, and that increasing scientific understanding is one of humanity's greatest endeavors. From the mid-1970s available funds for both teaching and research had fallen well short of demand, as the universities in the United Kingdom were caught between the remorseless sophistication of knowledge and their constrained resources.

In May 1980 the then-new Prime Minister, Mrs. Thatcher, said that the science budget would be protected against inflation and against the new Conservative Government's efforts to reduce public expenditure. Indeed in the following decade the science budget grew from about £500 million to more than £850 million per year. However, rhetoric in high places does not always translate into appropriate action, and other steps were taken that undermined the increase in science funding. At the end of 1980, the Secretary of State for Education and Science, the Government Minister then responsible for such matters, announced that the recurrent budgets of the universities were to be reduced by 13% over 3 years. Thereafter in political circles, when the allocation of money to the universities in "block grants" was discussed, it became common practice to use the buzzwords "level funding" to describe a situation in which public spending on higher education would not increase in real terms. This meant that the fraction of university spending for support of science fell short of that required to maintain the infrastructure of laboratories and to exploit new opportunities. Moreover, what was termed level funding proved illusory. Student numbers have increased, resulting in a very significant increase in teaching load for individual academics, thus reducing the time available to them for research. Also, the cost of research has risen more rapidly than general price rises because of the increasing sophistication of science. Furthermore, inflation was generally underestimated, particularly in relation to salary rises, over which the universities had little control, since in Britain these are negotiated nationally. Failure to fully index the funding of universities for the ravages of inflation during the 1980s has undermined their infrastructure. Less funding for the same system, indeed one now required to accept more students, has led to the appearance of the new phrase "efficiency gain" in directives from bureaucrats.

The U.K. higher education system has always been regarded as efficient in the sense that nine out of ten entering students graduate, as opposed to fewer than one-third in some European countries, and this has been regarded as its special strength. Britain produces almost as many graduates as France or Germany but for many years has admitted about half as many students to university. Tough selection resulted in low wastage, but this policy cannot be sustained on the edge of the 21st century. One should never underestimate how the United States has benefited because so many of its people have had at least some contact with higher education. Britain is moving into a similar system, but will have to live with a higher wastage rate of students prior to completion of their studies. British universities are educating 19% more students than 3 years ago, and growth in the new universities, previously designated as polytechnics, is also substantial. However, over the same period the money available to educate each student in Britain has fallen by about 13%. Although this is referred to by the politicians as an efficiency gain, for academe it represents a very serious depletion of the infrastructure for libraries, laboratories, building maintenance, and so forth.

For readers unfamiliar with the manner in which money is provided for support of the universities in the United Kingdom, it is necessary at this point to give a brief and therefore necessarily oversimplified explanation of a complex and rapidly changing arrangement. After the First World War, problems within the universities, mainly financial but some organizational in nature, resulted in Parliament assuming an ever-increasing responsibility for the greater part of the funding. The universities had become unable to generate the cash needed to meet the many demands upon them. This problem was solved via grants from Parliament to the system. The universities were able to continue to operate under their own individual charters because a good mechanism was introduced whereby they were protected from direct political intervention. This was accomplished by Parliament in 1919 setting up an organization called the "University Grants Committee", which had the task of advising the Government of the day as to the distribution of any grants made by Parliament and of consulting with the universities and other bodies on the preparation and execution of plans for development. Although the members of the UGC were chosen by the Government in power, membership was changed from time to time and consisted mainly of persons drawn from academia and industry.

This situation continued until and through the Second World War, and for several years thereafter, so that planning by individual universities was conducted in a sensible manner. Funds were earmarked by the UGC on a quinquennial basis, and universities knew within reasonable bounds what level of support from public funds they were likely to receive. In addition to these so-called "block grants" from the UGC, which amounted to about 48% of income in 1989, the universities receive additional money through student fees and from research grants disbursed by Research Councils, the roles of which are discussed later. For some years local district authorities in the United Kingdom have been required to pay the tuition fees of all students from their particular cities or districts qualified to enter a university. Thus a further ~14% of the income of universities is currently derived from this source, which is also dependent on taxation. Students may qualify for living costs, again paid by their local district authority, but these grants, unlike tuition fees, are means-tested against parental income. Hence with a further 20% of income obtained from the Research Councils, the university system in Britain has evolved during this century into one in which the funding is very largely (~80%) from the public purse. It is understandable, therefore, that politicians held responsible for raising money by taxation increasingly scrutinize the funding of higher education.

In recent times provision of money for the universities has become a relatively low voter priority, certainly ranking below that for schools and health care. Indeed, I recall one Minister saying to me that while I had presented a very good case to him for more money for universities, his colleagues would argue that there were fewer votes in increasing university funding than there were in providing more tax money for the escalating cost of hip replacements and other medical procedures under Britain's national health service, which employs in one way or another about one million people.

In 1989 the UGC was replaced by a new body, the Universities Funding Council (UFC). No plausible reason was ever given for this action, but it probably represented a desire on the part of Government to introduce a more manageable system whereby the universities would become more controlled from the center and be made more competitive among themselves in applying for whatever public funds the Treasury might assign for higher education. One suspects that the politicians and civil service mandarins thought that the UGC had become too user-friendly in regard to their support of universities. Ironically, after only a very short life, the UFC has been assimilated into yet another body (a Higher Education Funding Council) that has assumed responsibility for funding both the universities and the polytechnics in Britain, the latter having recently been redesignated as "universities". Fusion of the funding for higher education in this manner is likely to have major repercussions, since the unit cost of educating a student in a hitherto "polytechnic", although rising, is significantly less than that required in a university. Thus in 1988 Government spending averaged for all disciplines at £5719 (about $8600) per year for a university student and at £3353 (about $5000) per year for a student attending a polytechnic institute. This differential is now closing, as a result of the universities, with no significant increase in their resources, packing more students into existing classes and laboratories.

One can anticipate that any British Government, whatever its political hue, and wishing to reduce public expenditure, will be very strongly tempted to level university unit costs down to those of the institutions previously named as polytechnics, and which were originally created to fulfill a different role from that of universities. In another significant change in policy it is intended that the Scottish and Welsh systems of university education be run independently from that in England. By now a reader will appreciate that because of the very frequent changes and uncertainties in higher education in Great Britain in recent times, planning has reached the stage in some quarters where decisions are made on the assumption that the end of the world will happen before the next fiscal year arrives.

Science played a crucial role in the Second World War, and it was well appreciated that many scientists from academia had contributed to the eventual victory. For decades no one questioned the link between research and other scholarly activities in the universities. Moreover, certain agencies were given the responsibility for injecting research funds into the university system. These agencies, now termed Research Councils, have gone through several changes in name, as over the years their responsibilities have been redefined.

As with the UGC and the new HEFC, the Research Councils have their own annual budgets. Until very recently, when organizational changes were announced, which are referred to later, it was the overall responsibility of the Cabinet Minister mentioned earlier, the Secretary of State for Education and Science, to determine the allocation of the money between the various Research Councils. However, he or she delegated this responsibility to a junior Minister in the Department of Education and Science, and since this department naturally devotes more attention to school children than to scientists, the needs of the latter generally have a lower profile. The Secretary of State's decisions, made in consultation with his junior Minister, also a member of Parliament but not a member of the Cabinet, were based on advice received from an Advisory Board for the Research Councils (ABRC). Funding for expenditure on scientific research is constrained within the limits of the amount of money made available by the Treasury. The responsibility for support for chemistry research in the British universities lies primarily with the Science and Engineering Research Council (SERC), with a 1993–1994 budget of approximately £580 million ($870 million).

Until relatively recent times it was assumed that all senior scholars, even leading scientists, had a significant obligation to undergraduate teaching as well as research, and that all who held academic appointments were expected to do research and were given reasonable support to do so.[226] The base funding for the latter was provided by the UGC–UFC distributing money to the universities on the assumption that every academic staff member would spend approximately 40% of time on research activities, and so the block grant assigned to a particular university made allowance for this salary component. The amount of UFC money distributed in the fiscal year 1990–1991 to the university system for support of research was £830 million ($1250 million; I have chosen an exchange rate of $1.50 to the pound sterling) of which about £665 million ($1000 million) was intended to support research in laboratory-based subjects. This money was expected to be used by the 50 or so universities* to provide the necessary laboratory infrastructure, including for example payment of the salaries of technical staff and provision of the basic instrumentation one might expect to find in any science laboratory.

Until recently universities received a global sum (the block grant) for teaching and research to be employed as they choose, under the autonomy granted to them by their charters. Suspicions have arisen that in some institutions part of the money intended for research support has been diverted for other purposes, a point discussed further later. It was the responsibility of the Research Councils, mentioned earlier, to provide funds for state-of-the-art instrumentation, scholarships for graduate students, and grants to individuals for work of timeliness and promise.

This "dual support" mechanism, which involves channeling resources to support research by two routes [UGC (now HEFC) and the Research Councils], was seriously damaged by

---

* The number of U.K. "universities" depends on definition, including whether or not the constituent colleges of London University are regarded as individual institutions or part of a federal structure. Moreover, the recent decision, mentioned earlier, to encourage the polytechnics to redefine themselves has at a stroke increased the number of universities to about 85 and student enrollment to about 570,000. However, many universities are small by non-U.K. standards, some having only 2000–4000 students.

the ~13% cut in the UGC block grant to the universities, mentioned earlier, which operated over 3 years starting in 1980, and by other factors also mentioned previously, such as the increased cost of research.† However, the financial problems of British science began much earlier. The optimism created by the founding of new universities in the 1960s quickly evaporated as the then Labour Government grew discontented with the cost implications, especially the money required to sustain research at all universities. In a famous speech, the then Labour Party Minister responsible for education and science warned academic staff of the British universities that "the party was over". From the 1970s on, the test applied by Government to the academic research sector has been that of whether the science budget has increased annually to match inflation—and often it has not—rather than whether the budget has increased as quickly as the increase in the number of researchers and costs of facilities.

New results stream forth from laboratories with such speed, and are reported in the literature so fast, that money required to exploit all of the projects has become politically unavailable. Neither Britain nor the United States has successfully addressed this problem. Arguments based on a steady state for research support go against the ethos of science, the endless frontier. Scientists are reluctant to define priorities other than to defend and seek to extend their own area of endeavor. Nevertheless, as chemists we must not allow experts removed from our discipline to decide priorities through our failure to articulate the riches chemistry has brought to society by the past liberal and nonrestrictive policy of research support.

After 1980, as a climate of constraint in funding developed, not surprisingly the research component of the UGC–UFC budget came under attack from several quarters, including the Research Councils and those who managed them. These organizations were being financially squeezed for a variety of reasons and saw that it was to their advantage to per-

---

† With the notable exception of mass spectrometers and the superconducting magnets used in NMR spectrometers, most of the instrumentation (X-ray diffractometers, electronics for NMR and ESR equipment, computers, and so forth) required in a modern chemistry department is not made in Britain. The decline in purchasing power of the pound sterling with respect to other international currencies over many years has placed additional strain on the provision of the "well-found laboratory" in the U.K. universities.

suade Government Ministers to raid funds earmarked for the universities. It is argued that the money distributed to institutions by the block grant in support of research, rather than to individuals, is soft money disappearing into a "black hole" because it is not subjected to the same degree of peer review as the grants from the Councils themselves. While there may be a small element of truth in this criticism, the "dual support" arrangement, whereby core funding came from the UFC, and money for research at the cutting edge came from a Research Council, *provided an important degree of flexibility* in the use of shrinking resources. Science departments could argue the case for money for a well-founded laboratory within committees inside their respective universities, and colleagues in departments could work together in an efficient manner in deciding priorities, for example, whether to upgrade X-ray equipment for the laboratory or to purchase a spectrometer of one kind or another. Moreover, in my own department, as surely happened in others, some of the block grant money was used to support the research of young members of the academic staff without their having to convince an external committee that their work deserved funding. This flexibility became crucial as a situation developed where the projects of one's colleagues were rated highly for support by external peer review, but were not in practice funded by the SERC because of lack of money in the grant stream. In the United Kingdom this has become known as the $\alpha-$ unfunded syndrome, for which a special tie with an unfunded $\alpha-$ logo is available for those who wish to advertise their situation. Ludicrously two of Britain's Nobel laureates (George Porter and Geoff Wilkinson) would qualify for the tie. In the North American culture a T-shirt carrying an $\alpha-$, instead of a tie, would be more in tune!

Transfer of money to the Research Councils from that previously allocated by the UGC, and its successors the UFC and the HEFC, has now begun in Britain, and probably represents the thin edge of a very thick wedge, despite noises to the effect that some base funding will remain at the disposal of the universities for the general support of research in science. Although it has been stated that these transferred funds (£150 million from August 1992) will be returned to universities as overheads on grants to individuals, it remains to be proven that this will be to the advantage of the investigator. The SERC, responsible for funding frontier research in chemistry, will be under continued pressure to commit its money to the support of its large central facilities and to develop "special projects", all of which generate their own bureaucracies. It is not likely, therefore, that the number of individual grants will increase as a result of garnering money previously at the disposal of the UFC (now HEFC). Indeed, it is easier to reduce the number of single investigator grants than either to dismantle long-term commitments or to cut back on trendy projects designed to capture the attention of politicians and the public. Moreover, the opening of new facilities provides "photo opportunities" for Ministers, as well as some of our scientific brethren. Such publicity does not harm career advancement of those in either category.

University Chemistry—The Way Forward

The Subject Review Committee I was asked to chair was one of several organized by the University Grants Committee shortly before it demise. Besides chemistry,[227] other review committees included earth sciences and physics. As my colleagues and I continued our deliberations over several months, we were all realistically aware that we were operating in the very dynamic situation, referred to earlier, where university and science funding policy was subject to change as rapidly as the British weather. In good faith the UGC notionally earmarked funds from its budget to meet the costs of implementing the recommendations of the various subject committees it had set up for the core sciences. However, during this period the UGC had to resolve emergency situations, including help for some universities in a precarious financial state. Funds allocated for one need in the system were thus diverted to deal with near-bankruptcy situations in other parts. Even so, very few members of any of the various UGC subject committees would have guessed that the Government intended to do away with the UGC and replace it with the UFC, as described earlier.

Fortunately for those in the earth sciences, the recommendations of their subject review came out well before the UGC was abolished. As a result substantial resources were made available to universities to reorganize and improve the facilities for departments of geology and related subjects. The chemists, physicists, and others were less fortunate because their reviews were issued later, during the last year of a lame duck UGC. The new UFC became primarily concerned with devising new funding schemes, following to some degree hints from Ministers and others of what they might do to move towards a "market economy" for education. These schemes included, for example, having universities bid for a substantial proportion of their annual funding on the basis of unit costs for educating students in particular subjects, a procedure designed to introduce market forces into university financing. As almost anyone would have predicted this was a time-wasting exercise, since once the bids from the various universities were assembled, there was a remarkable unanimity between institutions as to what it cost to graduate, for example, a B.Sc. chemist. Heads of individual universities were naturally quite capable of comparing unit costs among themselves, with advice from departments, and they would be reluctant to underbid each other for a host of different reasons. The deregulation practices imposed on airlines and other industries did not carry over to the universities with any success, as any scholar would have predicted.

In this quagmire, not surprisingly, the UFC showed very little interest in the recommendations made by the chemists and the physicists in their respective reviews as to what might be done to nurture teaching and research in these important subjects in the U.K. universities. The chemistry review unequivocally stated, for those who cared to read, that our discipline is an essential and lively core subject, and that producing well-trained undergraduate and graduate students is vital for the survival of one of the United Kingdom's very few successful manufacturing industries. It was also stated that realistic funding was required to prevent further attrition of the overstretched resources of chemistry departments.

The UFC seemingly ignored the many recommendations contained in the review, including those stating the need to provide money to replace deficient basic equipment, to appoint younger staff to improve the age profiles of departments, and to reconstruct many chemistry departments to modern safety standards. However, as one looks back one can detect with a degree of satisfaction that the universities themselves have implemented some of the proposals, and the UFC (HEFC) is slowly proceeding to implement others while not admitting that they are doing so. The strong case presented for chemistry included a recommendation that departments should have at least three chairs, and in total 20 academic staff, with interests spread over different areas of the subject, in order to give its undergraduates a firsthand glimpse of the frontiers of the subject, to expose them to teachers with varied interests and styles, and to provide sufficient expertise and breadth within a department for the training of students to the Ph.D. level. University heads chose to interpret this somewhat controversial recommendation as implying that if this criterion were not met their institutions should not be deemed "universities". Consequently, since 1988[227] several vacant chairs in chemistry, left unfilled for budgetary reasons, have now been filled; still others have been created, and many senior staff have been promoted to personal chairs. Moreover, several younger staff have been appointed where previously administrators in universities were deaf to the pleas from department chairmen for staff replacements to fill frozen vacancies.

An unavoidable consequence of reorganization has been the closure of some dozen chemistry departments of the 55 or so existing in the U.K. system after university expansion in the 1960s. In a university system where funds are mostly provided from the public purse, it is in my opinion unrealistic to argue for the retention of departments that are unable to attract undergraduate students in sufficient numbers to justify the expenditure required for staff salaries. As mentioned earlier, some "universities" have fewer than 3000 undergraduates in all disciplines, and in several of these institutions the chemistry departments had been allocated insufficient resources to enable them to expose students to the variety of research themes necessary to provide a broad training. Fortunately, most staff in the chemistry departments that have been closed have been transferred to financially more viable institutions or taken into multidisciplinary units in their own universities.

The situation remains, however, far from static, as pressures increase to separate teaching in universities from research, and to create in the United Kingdom an elite group of universities, in order to concentrate limited resources. If this policy comes about, as seems certain, there will remain for several years many frustrated and demoralized academics in science departments in "universities" where research was once combined with teaching who will no longer receive Research Council support for research. Unfortunately, research has come to be identified with those themes favored by the Research Councils, and the ethos that research is an integral part of activities in a university has been weakened.

Many of the present difficulties have their origins in the expansion of the British university system in the 1960s, when it was accepted by the Government in power at that time, and their advisory body then (the UGC), that if the quality of teaching is to be of high class all academic staff must be closely in touch with research. Hence all the new universities were planned and initially funded on this premise. The opportunity to create several undergraduate institutions based on the model of the better American 4-year colleges was lost. Consequently, the British became locked into a system where "level funding" from the annual block grants from the UFC to universities meant in practice "underfunding", largely because of the rapid increase in running costs and the need to purchase increasingly sophisticated equipment for scientists to keep pace with developments. For chemists working in university laboratories this situation has been exacerbated by the large proportion of money committed in advance by the Science and Engineering Research Council to support "big science", carried out both in the United Kingdom and elsewhere as a consequence of long-term commitments to international collaborations.[228] Although funding for the Research Councils climbed by 14% in real terms between 1981 and 1989, this added funding brought relatively little benefit to the single investigator. The only comfort for the British chemist in academia in the present climate, and it is a cold one, is that their colleagues in other countries are beginning to encounter the same difficulties arising from a paucity of research funding. This state of affairs is unlikely to improve in the short or medium term because governments everywhere are now moving to reduce public expenditure.

The climate of university and science funding in Britain during the past 10 or so years, described earlier, was personally frustrating to me as it was to others. There were many things wrong with the system that could have been changed to increase efficiency, without the need to inject substantially more money. It is not so much the amount of money being spent on science in Great Britain, which is important, but the way it is spent, a feature referred to in the next section. A relatively small addition to the funding available for single investigators in university chemical laboratories would have relieved much frustration. The all too frequent changes in policy, the doses of rationalization and unsympathetic cost cutting, and the littering of the academic landscape with reports and surveys produced a time-wasting atmosphere for senior academics and seriously damaged the scientific careers of junior staff seeking research support. As the President of the British Academy has said, the great divide in the universities in the United Kingdom is no longer between the arts and the sciences but between the scholars and accountants.

As this book nears publication, the way in which public support for science is managed in Britain is undergoing major restructuring. A Minister in the Cabinet, rather than a junior Minister in the Department of Education and Science, has assumed responsibility for overseeing that portion of science supported by the Research Councils. The Research Councils are to be reorganized, with funding for particle physics and astronomy, and for biology, being removed from the ambit of the SERC. The SERC will thus become an Engineering and Physical Sciences Research Council, supposedly to provide a focused body that will stress technology transfer between the science base and industry. It is most unlikely that this reorganization will produce more money for basic research for academic staff in chemistry departments. To the contrary, with the United Kingdom carrying out only about 5% of the world's research, the new plans are unashamedly geared to underpin research areas deemed to be crucial to industry. The increased emphasis on applied science with less on curiosity-driven research, and the trend to separate teaching from research, bodes ill for chemistry in the U.K. universities.

Research Funding

Having lived during an exciting era for chemistry it is perhaps worthwhile to include some further comments on trends in research funding. It is a truism that scientific discoveries are not possible without adequate money for research. While at Harvard University my own work was supported by the U.S. Air Force Office of Scientific Research and the National Science Foundation, but after returning to Great Britain in 1962 my funding was mostly derived from the Science and Engineering Research Council and from my own university, both operating under the dual support system described earlier.

For many years grant renewals from the SERC, following peer review, depended on the proposals containing new ideas and on the previous track record of the investigator. Unfortunately, recent trends in science funding policy in Great Britain have severely inhibited timely and promising research in chemistry. Resources allocated to fundamental research have become dangerously inadequate.[227] The current problems besetting support for chemistry in the United Kingdom are gaining ground in several other countries in Western Europe and are also appearing to an increasing degree in the United States. I refer to the disproportionate amount of money being directed towards so-called strategic rather than "blue-skies" research and the ever-increasing proportion of the budget for the core sciences being consumed by large central laboratory facilities.

Central facilities, once established, are virtually impossible to close, or even reduce in size. Yet, as science advances, the particular services provided may no longer be required, may have been duplicated elsewhere, or may have become so expensive that if support is to be continued at what is deemed the required level, then funding in other science areas has to be reduced. The easiest source of extra money to patch over difficulties when budgetary problems arise is to "borrow" from the funds earmarked for grants to single investigators working at universities. This procedure has become the all-too-ready soft option adopted by the SERC to solve its financial difficulties during the past decade, a policy resulting in an increasing number of α-rated peer-reviewed grant applications for individual investigators being unfunded.

Strong defensive mechanisms are set up to justify not closing any large facility with several hundred employees who hold quasi-academic and civil service appointments. Closure requires terminating the appointments of directors, associate directors, project managers, administrators, secretaries, scientists, technicians, as well as a host of other staff, down to those people required to maintain, clean, and secure the buildings and look after the car parks. One defensive posture adopted is to hold out the hope of selling instrument time on a national facility to industry or to the scientific community of another country that does not have, for example, a suitable neutron source or large telescope. Having become committed to what is essentially a public works project, using science as its legitimation, the advocates of these facilities discover that funds are insufficient to sustain the enterprise. While I was a member of several committees, during both the Thatcher and earlier years, I found it painful to listen to the verbal reports of senior scientific staff from central facilities describing their at best partially successful entrepreneurial activities designed to keep a large facility alive through the window dressing of foreign participation. However, one's concern about the efforts made to sustain a large facility were quickly tempered by the knowledge of the frustrations and loss of morale experienced by young colleagues in university chemistry laboratories receiving letters from the SERC with news that yet another of their applications, after being peer reviewed, had been placed in the α category, but had not been funded. Not surprisingly, Ministers tend to be unsympathetic to the situation, since they can point to the 14% increase in funding for the Research Councils during the period 1981–1989, mentioned earlier. They adopt the reasonable posture of saying that the remedy is all in the hands of the scientists who are members of the relevant committees deciding priorities in spending among the various branches of science and assert, rightly, that politicians should not set priorities.

In the competition for scarce resources for curiosity-driven research, chemistry in Britain has also suffered badly as a result of the diversion of funds away from the grant stream into research "initiatives" and interdisciplinary research centers. This situation is supposed to produce results having potential industrial applications "in the short and medium term", to quote SERC documents that crossed my desk almost every week during the 1980s. There is, however, no evidence that the direction of support into large interdisciplinary research centers, a Stalinist approach, will bring any greater economic benefit or lead to a better use of public money than the previous "free market system" of support based on grants to individuals. Indeed, all evidence points in the opposite direction. The constant drumbeat about the lack of relevance of academic research and its perceived failure to produce a steady stream of usable inventions has had a debilitating effect. There are scientists who think that the best strategy for extracting new money for research is to identify for politicians areas where basic discoveries are thought likely to lead to economic gains in a relatively short time span. This approach serves as a palliative to politicians who have become disillusioned by the fact that investment in basic research in universities does not result in an increase in national wealth on the political time scale, that is, the time period that elapses between elections. The wisdom of this approach by colleagues to gaining extra funds is, however, questionable. Fortunately, and predictably, the original enthusiasm in the United Kingdom for interdisciplinary research centers has evaporated as Government circles have come to realize that these centers might do near-market research for industry on the cheap at the taxpayers' expense, to the detriment of basic science in the universities. Moreover, some of the interdisciplinary research centers have not attracted matching funds from industry to the degree that was hoped by the bureaucrats, which is perhaps no surprise as industry seeks high-quality, broadly trained chemistry graduates rather than those locked into a specific interdisciplinary area.

Unfortunately, those responsible for reorganizing British science these days, mostly physicists and engineers, have failed to appreciate that our core science chemistry remains a highly competitive activity that depends on the initiative and personal creativity of individuals generally working in small groups.[229] It is dangerous to extrapolate from one's own experience in research to other fields and to assume that a research policy that is suitable for physicists or engineers is equally well suited for chemistry. Given a "well-founded" laboratory many chemists are able to do innovative and competitive research to the highest international standards with the aid of relatively small grants. Indeed, Mrs. Thatcher,[230] whose first profession was chemistry, once said, "We should be ready to support those teams, however small, which can demonstrate the intellectual flair and leadership which is driven by intense curiosity and dedication". Unfortunately this free market message has not been taken to heart by those who have made U.K. science research policy in recent times, and the SERC, which is primarily responsible for responsive mode funding in chemistry via peer review, has become psychologically incapable of distributing its non-earmarked money for single investigators. Consequently, as indicated earlier, far too much good work does not receive support. As the all too frequent crises in science funding arise, the money for the grant stream to support individual investigators, to provide state-of-the-art equipment, and to fund research students, is reduced or at best deferred for several months. It is the very pattern of support that has developed, even more than the scale of the available money, that has created the impossible pressures on responsive mode funding for investigators in small science in the British universities.[228] Within the domain of "small science", in addition to chemistry, one should include substantial parts of biology and physics.

During the past 40 years, success in chemistry in the West, especially the United States, stemmed from the diversity of chemical research, its multitude of funding sources, and the absence of a centralized encumbering directorate. It must be appreciated, however, that although society generally looks with favor on scientific endeavor, public funds are not limitless. The argument one hears from some in the academic community— that all capable scientists deserve to have their research supported from public funds—is damaging to university science at a time when there is too much emphasis on obtaining research results for career advancement. Moreover, these same public funds, drawn from the taxpayer, are required for health care, for primary and secondary education, and for many other needs linked to the quality of life. It can be argued that much scientific research benefits society and should be supported out of the public purse, but the extent to which this support justifies the activities of every scientist now in academia, when every experimental result usually suggests several other projects worth funding, is another matter. It is very unlikely that the well-funded system existing in the U.K. universities in the 1960s and 1970s will be supported by any future Government of whatever political hue.

Channeling limited resources into fewer areas appears to meet the criticism that scientific knowledge is being acquired at a faster rate than it can be usefully assimilated, and that much of the data acquired do not represent a significant scientific advance. Although one may have some sympathy with these views, the links between discoveries in basic science, exploitability, and wealth creation are so complex that expensive errors are in the making. Advances in chemistry are small, many workers contribute, and it is the cumulative impact of their discoveries that continues to transform chemistry and the world. Nevertheless chemists, like those of other disciplines, shy away from addressing the problem of what proportion of a country's limited resources should be devoted to pure science. We are in an era in which it is not possible to fund every good project. A degree of selectivity and collaboration in research is thus inevitable, but selectivity of support by mechanisms involving a narrowing of the discipline reduces the research base of any country and endangers its future prosperity and even its culture.

Frustrating as it may be to those responsible for funding, the best basic research is done by individuals pursuing their own interests and naturally being drawn in directions they consider important. *Identification of relevant areas for so-called basic research by a committee, the members and secretariat of which are often divorced from active research, or in some cases have contributed little of significance, leads increasingly to the formation of self-interested and self-serving subgroups of people who attempt to mold their work to the aim of the initiative.* The result has been a chilling effect on chemistry in the universities in Britain. Moreover, such a policy leads to the submission of cosmetic proposals by investigators bidding for funds to work in areas for which they lack an intellectual commitment. The idea of "pork-barrel" politics will be familiar to U.S. readers. Competition between groups is becoming measured more in economic terms than in synergistic rivalry between scholars. Unfettered research in universities is vital to any country on two main counts. First, it is a source of the well-trained scientist that industry needs to progress and to adapt in response to changes in the economic, technological, and social climate. Second, universities play a critical role in providing the basic knowledge on which humanity depends for an increased quality of life. Insistence on proven relevance for basic research, surely orthogonal concepts, leads to increasingly costly and cumbersome administrative and bureaucratic controls to the detriment both of science and the taxpayer. Curiosity-driven research is an indispensable part of a complete national research strategy.

The SERC approach to the support of chemistry at universities in Britain has in the last decade been at best misguided, with funding following schemes generated by desk workers who think they can define the patterns of our research. The general policy has been to blend chemistry into various other areas so that special topics avoiding the name "chemistry" are created and "hyped up". Core chemistry has been neglected, and in order to obtain support, chemists have redesignated themselves as material scientists or bio-this or bio-that. Lost has been the recognition of the empiricism of chemistry that led to the discovery of ferrocene, the platinum drugs, C_{60}, and a host of other important compounds. A plan to close all research groups with the word "Center" or "Institute" in their title would do much to preserve research in chemistry in the British universities, as would the transfer of most of the funds currently used by the SERC (now apparently to be the Engineering and Physical Sciences Research Council) for its in-house facilities to the universities for research initiated by the scientists themselves.

Visibility

With the competition for research funds becoming harder each year, attempts are being made by various organizations to quantify the track records of individuals and departments. Although most chemists would agree that published work should weigh heavily in any evaluation, any assessment that depends entirely on publications is hazardous. Moreover, there are dangers if such data are accepted as reliable measures of scientific competence merely because they are presented in a numerical manner.

According to the Institute of Scientific Information (ISI), *Angewandte Chemie, International Edition in English* has an impact factor higher than that of most other journals. This is not surprising in view of its excellent coverage presented in a most attractive format for readers. In late 1988 I received a letter from the editor, Peter Gölitz, pointing out that an article I had published in his journal in 1984 had, according to *Science Citation Index*, been the most cited chemistry article for that year over the period 1984 through 1986. However, while it is always a pleasure to have a letter pointing out that others are interested in your research, it is essential to treat information of this kind

Relaxation at the general assembly of the Gesellschaft Deutscher Chemiker, Heidelberg, 1985. Left to right, Judy Stone, Drs. Peter Gölitz and Eva Wille of Angewandte Chemie, *and me.*

with some caution, particularly if it refers to the number of publications instead of citations per paper.

In an article in a January 1992 issue of *The Economist*, entitled "Pity the Poor Typist" and using data from the ISI, mention was made of some 20 scientists who had published 9365 papers between 1981 and 1990, getting their names into print, on average, every 9 hours and 20 minutes. Perhaps not surprisingly a crystallographer with 948 articles headed the list, being a coauthor every 3.9 days! It is not of course how much one publishes but who cares. Even combining numbers of papers with citations per paper is open to misinterpretation, especially if the data are averaged for groups of chemists in one location. In March 1992 the British newspaper *The Daily Telegraph*, employing ISI data, listed several university chemistry departments in Great Britain in order of citations per paper over the period 1984–1990. The first three departments (by number of papers/average citations per paper) were Cambridge (1479/11.28), Bristol (792/11.04), and Oxford (966/10.18). The newspaper's science editor wrote that the information would be used by the Higher Education

Funding Council in making decisions on how to allocate research funding. I am sure that this would be but one of several criteria used by the HEFC for distributing limited resources, but even so the data are open to misinterpretation on several grounds. For example, one highly cited paper by one author serves as cover for the inactivity of colleagues in the same department; also, some areas of a department will be more research-active than others. The article went on to say that in terms of worldwide impact Cambridge was 35th, with the top 10 being American. Although probably few would dispute that the top 10 are in the United States, to imply from such data that there are 34 chemistry departments superior to Cambridge is laughable. Unfortunately, when the ISI data reach the media all manner of obfuscation and ill-judged comments appear. For example, the science editor of *The Daily Telegraph* refers to the 1521 papers "churned out" by the chemistry department of the University of Texas (Austin). There is no reference to the fact that the size of a department will influence the amount of published work, while inclusion of the phrase "churned out" implies some factory producing a product of perhaps little real significance. Yet a cursory examination of the ISI study, which threw up one of my 1984 papers as being the most highly cited in chemistry for that year, revealed that five papers from groups at the University of Texas at Austin were in the top 100, which is not a bad batting average for one department.

Not surprisingly, several of the contributors to this series, Profiles, Pathways, and Dreams, are ranked highly in *Science Citation Index* both for numbers of papers and citations per paper. It would be remarkable if they were not. Moreover, I suspect that many, like me, make an effort to publish a paper or two coauthored with every person who works in their respective groups, in order to make the coauthor's curriculum vitae look reasonable for the next job application. Occasionally this may be impossible, but I am pleased to say that I have had few young colleagues with whom I have not coauthored a paper.

From Bristol to Texas

In 1990 I moved from Bristol to Baylor University to continue my researches under the auspices of the Robert A. Welch Foundation. This move came about through my having reached the mandatory retirement age of 65 in the British university system. Generally those who retire from a chair in a British university are not in a position to continue their research, since both space and funding are at a premium. Since I was not prepared to accept such a rigid and unwelcome system, which assumes that at a certain age one can contribute no more to science, a move to the more relaxed academic environment operating in Texas was impossible to resist. My enthusiasm for synthesizing organometallic compounds remains undiminished after 40 years. Selby Knox informs me that I always regard the latest organometallic molecule synthesized in my group as being the most exciting.

I express my thanks to my new colleagues at Baylor who have welcomed my wife and me to our new habitat. Special thanks are due to Dr. John S. Belew, until recently Provost and Vice-President for Academic Affairs and now Jo Murphy Professor of International Education. An organic chemist at heart, John Belew organized my transition to Texas in an expeditious manner. By retaining a home in Britain, as well as having one in Waco, Judy and I can avoid both the British winter and the Texas summer. It would be nice if Dallas—Fort Worth airport were closer to Gatwick airport, Waco closer to Dallas—Fort Worth, and Gatwick closer to Bristol, but travel between two

A farewell at the Palace before leaving for Texas.

homes is a minor inconvenience when compared with the benefits of exposure to two very different cultures.

I had held the Chair of Inorganic Chemistry at Bristol for
27 years and obviously had some regrets on leaving. However,
it was in many respects a good time to depart, because of the

growing infiltration into daily life of bureaucratic affairs, leading within the university system to the over-organization of scientific work both in teaching and research. Far too much of my energy was being expended on completing questionnaires of one kind or another issued by the university administration or outside bodies, seeking funds, and defending the needs of the staff of the department at meetings held both at the local and national level.

Four Bristol graduate students of mine, who had not completed their Ph.D. studies at the time of my "retirement", have enjoyed the opportunity to commute back and forth between Texas and England, both to continue their experimental work and to write up their results. Several postdoctorals have joined the group, and my special thanks go to Nick Carr for coming to Baylor at the same time as I and setting up the laboratory with the necessary equipment. I plan to have only postdoctoral assistants at Baylor, because Ph.D. work in the United States takes too long to complete. American chemists need to address this problem. The mean registered time to the doctorate from completion of the bachelor's degree has risen to 5.8 years, to be compared with about 3.5 years in Britain. Therefore, if I were to take on graduate students they would still require mentoring when I will be perhaps too old to devote sufficient time and energy to their welfare. Postdoctoral colleagues are in a different category. One can be more relaxed and enjoy collaborating on more speculative projects, without worrying too much about generating thesis material.

Research now continues in Texas, and my group is presently focusing on the chemistry of alkylidyne(carborane)—metal species such as those indicated in formulae **103–105**. What is exciting and unusual about these new alkylidyne-(carborane)—metal species, and how did we come to prepare and study them? All chemists build on the discoveries of others, and good research ideas often stem from a fusion of seemingly unrelated work carried out in widely separated laboratories. This manner of thinking led me to consider the interesting chemistry that might result from a blending together of the alkylidyne metal chemistry of Fischer with the metallacarborane chemistry of Hawthorne. A successful synthetic chemist must have a good knowledge of the chemical literature in order to put this approach into practice, but most read widely. The species

103–105 display reactivity patterns that combine in an interesting way those shown by CR fragments with those shown by the 12- and 13-vertex cages MC_2B_9 and MC_2B_{10}, respectively. Addition of coordinatively unsaturated metal–ligand groups to the salts **103–105** affords cluster compounds (Scheme VII).[231] In many of the structures the cages play a noninnocent role, since as well as being η^5-coordinated to tungsten they form exopolyhedral B–H $\rightharpoonup$ M or B–M linkages to adjacent metal centers. I am indebted to several excellent co-workers both at Bristol and at Baylor for developing this field, including Franz-Erich Baumann, Steve Brew, Nick Carr, David Devore, Steve Dossett, Alun James, Alisdair Jelfs, Sihai Li, and Max Pilotti.

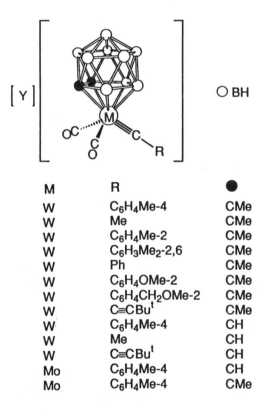

M	R	●
W	C_6H_4Me-4	CMe
W	Me	CMe
W	C_6H_4Me-2	CMe
W	$C_6H_3Me_2$-2,6	CMe
W	Ph	CMe
W	C_6H_4OMe-2	CMe
W	$C_6H_4CH_2OMe$-2	CMe
W	$C\equiv CBu^t$	CMe
W	C_6H_4Me-4	CH
W	Me	CH
W	$C\equiv CBu^t$	CH
Mo	C_6H_4Me-4	CH
Mo	C_6H_4Me-4	CMe

Y = NEt_4, PPh_4, $AsPh_4$, NMe_3Ph, $N(PPh_3)_2$, etc.

103

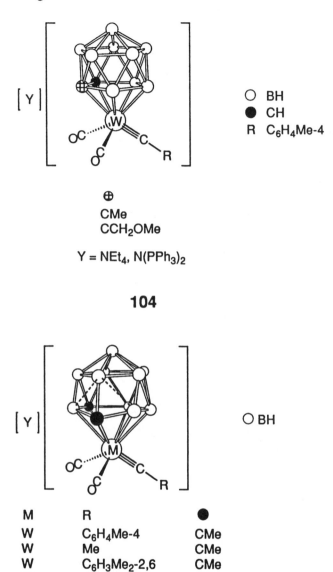

○ BH
● CH
R C₆H₄Me-4

⊕
CMe
CCH₂OMe

Y = NEt₄, N(PPh₃)₂

104

○ BH

M	R	●
W	C₆H₄Me-4	CMe
W	Me	CMe
W	C₆H₃Me₂-2,6	CMe
W	C₆H₄Me-4	CH
W	Me	CH
Mo	C₆H₄Me-4	CMe

Y = NEt₄, N(PPh₃)₂, NMe₃Ph

105

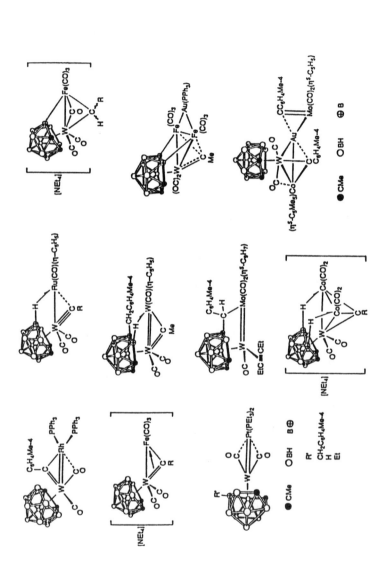

Scheme VII. Compounds with metal–metal bonds derived from use of the reagents $[X][W(\equiv CR)(CO)_2\text{-}$
$(\eta^5\text{-}C_2B_9H_9Me_2)]$ *[103, X is* NEt_4*,* PPh_4*,* $N(PPh_3)_2$*, and so forth] in synthesis. In several of the products the carborane cage adopts a nonspectator role as a ligand, a feature thereby giving impetus to this field of study.*

Dr. David Devore, a postdoctoral worker from Kansas State University, discussing results in my office at Bristol in 1987.

The results obtained by treating the salts **103–105** with electrophilic reagents in the presence of donor molecules are of increasing interest. The addition of protonic reagents to the salts would be expected to yield initially neutral alkylidene species, for example, $[M(=CHR)(CO)_2(\eta^x\text{-}C_2B_nH_nR'_2)]$ ($x = 5$ and $n = 9$ or $x = 6$ and $n = 10$; R' is H or Me). However, these products could be anticipated to react further because the metal centers are electronically unsaturated and should capture a donor molecule. Moreover, the alkylidene fragment lies on the surface of an icosahedral cage with neighboring and reactive B–H groups, and therefore insertion reactions are likely. Scheme VIII depicts the result of protonating the species **103** (M is W, R' is Me, and R is $C_6H_4Me\text{-}4$) with $HBF_4 \cdot Et_2O$, or with aqueous HCl at $-78\ ^\circ C$ in the presence of CO. Both acids give

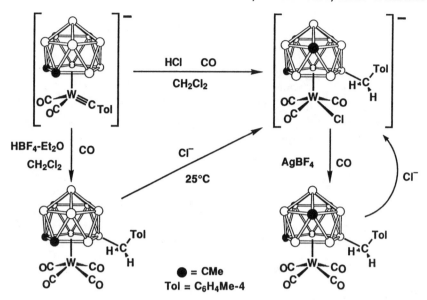

Scheme VIII. Protonation of the salts [X][W(≡CC₆H₄Me-4)(CO)₂-(η⁵-C₂B₉H₉Me₂)] (X is NEt₄ or PPh₄) with HBF₄·Et₂O or HCl in the presence of CO.

products apparently resulting from insertion of an initially formed CHR moiety into an adjacent BH bond, but with HCl the product is another salt in which the cage carbon atoms are no longer connected. Protonation with $HBF_4 \cdot Et_2O$ in the presence of the molecules $[M(\equiv CR)(CO)_2(\eta^5\text{-}C_5H_5)]$ (M is W or Mo) affords dimetal compounds such as the molybdenum–tungsten compound shown in Scheme IX. However, protonation with HX (X is Cl or I) followed by $[Mo(\equiv CC_6H_4Me\text{-}4)(CO)_2(\eta^5\text{-}C_5H_5)]$ and $AgBF_4$ to remove X as AgX gives another dimetal species that is a polytopal isomer of the product obtained via $HBF_4 \cdot Et_2O$.

That alkylidene–tungsten or –molybdenum species are the initial products in these protonation reactions is supported by results obtained when these processes are carried out in the presence of certain chelating phosphines. As depicted in Scheme X, it is possible in some instances to capture the alkylidene ligand as an ylid, thus preventing its insertion into the cage B–H group.

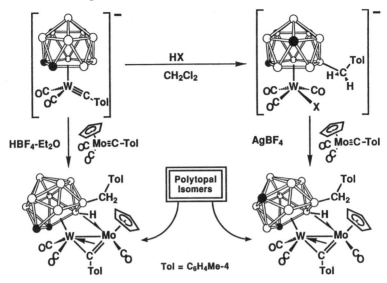

Scheme IX. Protonation of the salt 103 (M is W and R is C_6H_4Me-4; cage CMe groups) with $HBF_4 \cdot Et_2O$ in the presence of [Mo-($\equiv CC_6H_4Me$-4)$(CO)_2(\eta^5-C_5H_5)$] and the synthesis of a polytopal isomer via the use of HX (X is Cl or I) and the same molybdenum reagent with $AgBF_4$.

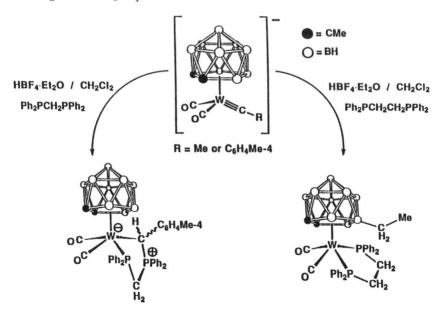

Scheme X. Protonation of the salts 103 (with cage CMe groups) in the presence of chelating phosphines.

Particularly intriguing results are obtained with the introduction of active groups into the alkylidyne substituent. Thus protonation of the salt [NEt$_4$][W($\equiv$CC$_6$H$_4$CH$_2$OMe-2)(CO)$_2$(η^5-C$_2$B$_9$H$_9$Me$_2$)] with HBF$_4$·Et$_2$O or HI (aqueous) yields the different products shown in Scheme XI. Elucidation of the pathways by which such complexes are formed remains an exciting challenge, as does the study of reactions of the kind shown in Scheme XII, where either CH(C$_6$H$_4$Me-4), or BH and CH(C$_6$H$_4$Me-4), fragments are eliminated, in the latter case with collapse of a 13-vertex WC$_2$B$_{10}$ cage into the 12-vertex WC$_2$B$_9$ icosahedral structure. This work raises numerous possibilities for further studies, as should all research in the area of chemical synthesis.

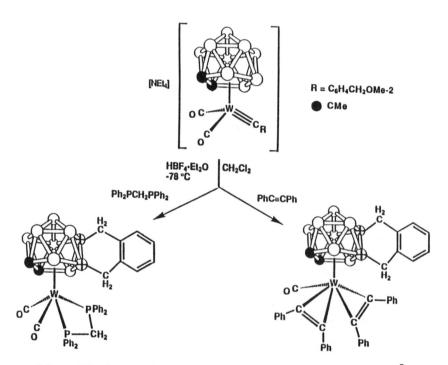

Scheme XI. Protonation of [NEt$_4$][W($\equiv$CC$_6$H$_4$CH$_2$OMe-2)(CO)$_2$(η^5-C$_2$B$_9$H$_9$Me$_2$)].

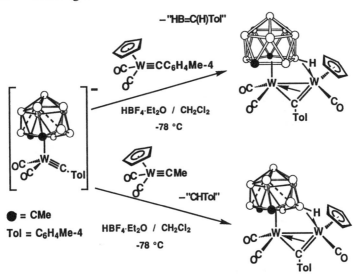

Scheme XII. Protonation of the salt [NEt$_4$][W≡CC$_6$H$_4$Me-4)(CO)$_2$-(η^6-C$_2$B$_{10}$H$_{10}$Me$_2$)] in the presence of the complexes [W(≡CR)(CO)$_2$-(η^5-C$_5$H$_5$)] (R is Me or C$_6$H$_4$Me-4).

Why Make New Compounds?

I have often been asked why I like to make new organometallic compounds and whether any of those reported by me and my co-workers are useful or significant. Chemists make all kinds of excuses for their activities that lead to discovery, but those engaged primarily in synthesis are probably mainly drawn by the lure of molecules with unknown structures likely to display new patterns of reactivity. One cannot deny the aesthetic appeal of new molecules. The building of our discipline as a science has as its foundations the acquisition of a huge dividend from chemical syntheses. With the elapse of time the original discoverers of a compound are usually forgotten. Moreover, the usefulness of a particular molecule is often hidden for many years after its discovery, until workers in other areas use the compound to sharpen principles, carry out creative further syntheses, or advance theory. Two examples from my own work, the synthesis of the molecules $Hg(CH=CH_2)_2$ and $B(C_6F_5)_3$, suffice to illustrate this theme.

John Pople and his co-workers have developed a general theoretical procedure, based on ab initio molecular orbital theory, for the computation of the total energies of molecules at their equilibrium geometries. One such study[232] involved ethyne, ethene, and both the vinyl radical and cation. Such a predictive procedure, able to reproduce experimental data accurately, is clearly of very great significance to chemists. The theory predicts a bond dissociation energy for C_2H_4 [$D_0(C-H)$]

of 110.2 kcal/mol, in excellent agreement with experimental data (107–110 kcal/mol) obtained from photoionization studies on vinyl radicals generated by the reaction of F atoms with C_2H_4 and by the thermolysis of $Hg(CH=CH_2)_2$.[233] The latter compound was first made in 1957 in my group at Harvard by Bodo Bartocha, shortly after Dietmar Seyferth had commenced his pioneering studies on the chemistry of vinylmetal compounds referred to earlier.[32] Vinylmetal species were at the time new and potentially interesting, and so Bodo Bartocha prepared divinylmercury[234] with the intention of employing this reagent for further syntheses. The molecule has been for many years of little interest, and yet it has recently proved very useful for spectroscopic work highly relevant to important theoretical studies.

When I arrived at Queen Mary College London in 1962, A. G. Massey joined the academic staff of the college at the same time, at my instigation in order to strengthen inorganic chemistry in the department. As mentioned earlier in this book, some of our initial research on arrival in London was successfully directed, in collaboration with Paul Treichel, Peter Jolly, and others, towards synthesizing the first C_6F_5 complexes of the transition elements.[104] Alan Massey was interested in boron chemistry, as I had been for many years, and it was therefore natural that following my earlier discovery of perfluorovinylboron compounds, such as $B(CF=CF_2)_3$ with Stafford,[235] that I suggested that we prepare $B(C_6F_5)_3$. At that time the literature was littered with reports of attempts to make compounds with fluorocarbon groups attached to tricoordinate boron, but most of these syntheses had been unsuccessful because of the inherent tendency of such species to eliminate the thermodynamically very stable BF_3 molecule. Tris(pentafluorophenyl)-boron was therefore prepared by us, and its anticipated strong Lewis acid properties were demonstrated.[236] For example, with LiC_6F_5, the salt $Li[B(C_6F_5)_4]$ was formed, and adducts such as $Ph_3PB(C_6F_5)_3$ were characterized. Subsequently, the compound $B(C_6F_5)_3$ lay dormant in the chemical literature until 1991, when Tobin Marks and his co-workers reported[237] elegant studies showing that the compound could be used to abstract methyl groups from $[ZrMe_2(\eta^5\text{-}C_5R_5)_2]$ (R is H or Me) to afford highly active "cation-like" alkene polymerization catalysts $[ZrMe(\eta^5\text{-}C_5R_5)_2][BMe(C_6F_5)_3]$. Thus the answer to the question raised at the beginning of this section is that we make new compounds because we enjoy doing so. One hopes that the significance of a few will be immediately apparent, but progress in chemistry is such that, not infrequently, minor discoveries such as those of $Hg(CH=CH_2)_2$ and $B(C_6F_5)_3$ provide the means for other chemists to make important advances at a later stage. In view of the multitude of new compounds reported each year I am not surprised to read many articles mentioning species first made in my group but without any reference to our early work.

Epilogue

The successful careers of my co-workers, who have worked with me in research either as undergraduate or graduate students or at the postdoctoral level, is especially pleasing to me. To date, they are approximately 200 in number and have come from many countries, and some 70 currently hold academic appointments in universities throughout the world. It is obvious that without the help of these young associates it would not have been possible to carry out the research described in this volume, and more fully in some 700 journal articles, excluding reviews. My co-workers have suggested many original projects for research that I would not have conceived myself. Relatively few have been mentioned by name in this book, in part because I have focused only upon certain areas of my research, but they have all contributed positively both to the chemistry and to the happy, stimulating atmosphere of the laboratory. I would also wish to pay tribute to Mrs. Beryl Staves, my secretary at Bristol for 20 years. Numerous graduate students, postdoctoral assistants, and I owe her a great debt for assisting us in many diverse ways.

In my opinion high-quality teaching and research are inexorably linked. This idea was promoted at Bristol by the requirement that all bachelor degree students in chemistry in their final year of studies (third year in Britain) engage in a research project. Instead of receiving formal laboratory instruction, students join a research group where they were able to col-

With my wife on the occasion of my 60th birthday party. A model of
[Pt(C$_2$H$_4$)$_3$] was presented to me by my co-workers and is shown in
the photograph.

laborate with more experienced graduate students and postdoctoral assistants. The great majority of undergraduates were
greatly stimulated by having this opportunity to work at the
cutting edge of the subject. Moreover, by association with more
experienced workers they quickly learned essential experimental
techniques, such as manipulating sensitive compounds in the
absence of air and growing crystals. Research projects at the
undergraduate level serve the useful purpose of stressing that
chemistry is an intensely practical subject and that theory is
driven by experiment and not, as often presented, the other way
around. Usually, sufficient results were obtained that, when
combined with the more substantial work accomplished by others, could be included in publications. For this reason many of
the primary journal articles originating from the various inorganic chemistry groups at Bristol were multiauthored. I was
often asked by Americans why our papers carried so many
authors' names. I make no apology for this. Even though
undergraduate researchers generally contributed significantly

Twenty-seven years at Bristol are celebrated at a party at the home of Drs. David and Judith Howard in May 1990. The top photo shows me with Professor Selby Knox on my left and Professor John Spencer on my right. The bottom photo shows Judy Stone, Tricia Boag, me, and Professor Peter Maitlis, who was a member of my group at Harvard.

less to a project than their more experienced collaborators, the possibility of coauthorship strongly encouraged a team approach to research and greatly increased the enthusiasm of all the students. Of course undergraduate research projects are common in many universities in Europe and in North America. However, within Britain the procedure followed at Bristol was not typical. Students worked on one project under one supervisor for about 7 months, enabling them to follow up an idea in some depth, rather than spreading themselves over two projects in more than one field of chemistry during a limited period.

During my career at Bristol University, the inorganic chemistry section of the school of chemistry was greatly stimulated by study visits from distinguished colleagues, who spent periods of a few months to 1 year with us. Among these I must list four persons who made especially significant contributions to our research: Bob Angelici (Iowa State University), Martin Bennett (Australian National University), Bill Cullen (University of British Columbia), and Warren Roper (University of Auckland). In addition, I made time for and allocated resources for a lively visiting speaker program. Very few of the distinguished inorganic chemists who came to Great Britain escaped a visit to Bristol. In this program we were pleased to extend hospitality to, among others, W. Beck, H. Bock, E. O. Fischer, W. A. Herrmann, H. Nöth, H. Schmidbaur, H. W. Roesky, W. Siebert, H. Werner, and G. Wilke from Germany; F. Bonati, P. Chini, L. Malatesta, and R. Ugo from Italy; R. Usón from Spain; H. R. Allcock, N. Bartlett, F. A. Cotton, A. H. Cowley, L. Dahl, G. L. Geoffroy, M. F. Hawthorne, R. Hoffmann, J. A. Ibers, A. G. MacDiarmid, E. L. Muetterties, and D. F. Shriver from the United States; S. Otsuka and A. Yamamoto from Japan; and B. O. West and A. M. Sargeson from Australia.

The growth of organometallic chemistry in the past 30 years has been phenomenal, particularly in that part of the subject involving the transition elements. There have been several reasons for the rapid expansion. In the early years, following recognition of the sandwich structure of ferrocene, investigators trained in the classical divisions of inorganic, organic, and physical chemistry entered the field. Within a decade strong research groups were founded in several countries, and these recruited high-quality graduate students and postdoctoral assistants who were subsequently in great demand for appointments in indus-

INSTITUT FÜR ANORGANISCHE CHEMIE
DER UNIVERSITÄT WÜRZBURG
Prof. Dr. H. Werner

8700 WÜRZBURG, May 19. 1985
Am Hubland
Telefon (09 31) 8 88 - 261

Professor F.G.A. Stone

Dear Gordon,

Firstly may I give you my cordial congratulations for your 60th birthday, accompanied with best wishes for the future. It is hard to believe that you now step into your seventh decade as your activity and readiness to discover new terra incognita on the landscape of organometallic chemistry seems to be unbroken.

If I say activity, this word in a personal sense seems to me strictly connected with your name. When we met for the last time in Bristol in May 1979, after a busy day we had a fine dinner in one of the well-known restaurants near to the Bristol harbour. When we came back to your home, we immediately went into your study to listen to the news and particularly to the results of the general election. As far as I remember, this was at 9 p.m. While watching the TV news you gave me a bottle of wine (from France not Germany), but you took to your well-known writing-desk to look through various new journals and to correct the proofs of several of your manuscripts. After a while I fell asleep, awoke after midnight, to find a smiling host (because Mrs. Thatcher had won) and learnt while drinking the rest of the wine that all the work had been done.

Again congratulations - and keep as you are. Please, give my best regards also to Judy.

Yours sincerely,

Near the time of my 60th birthday I received many letters from friends, including the one reproduced here from Helmut Werner. The letter nicely captures the atmosphere of the manner in which we entertained visiting speakers, the majority of whom stayed in our home. Not all visitors, however, were in Bristol on the day of a British election.

German colleagues have played a major role in the discoveries made in organometallic chemistry during the development of this area in the past 40 years. This reproduction of Gottfried Huttner's amusing birthday greetings, sent from Konstanz in 1985, allows me to acknowledge my very good fortune in establishing friendships with most of the active German practitioners in the organometallic field. Gottfried Huttner has moved from Konstanz to Heidelberg and currently occupies a chair in chemistry at the ancient university there. Research in chemistry has enabled me and my wife to establish an extensive but nevertheless tightly knit network of human relations with other chemists, and in many cases with their families also. This network, the excitement of discovery, and the game of making new compounds, are probably the main factors for the word "retirement" not being found in my vocabulary. One is singularly fortunate if one is able to focus for most of one's waking hours on a type of work that grows into a hobby, while at the same time opportunities occur for social contact with colleagues of many different nationalities and ethnic backgrounds, all having in common an interest in investigating new fields.

Visiting speakers who came to my laboratory in Bristol were usually asked to write their signatures in colloidal gold on a porcelain tile; the tile was subsequently heated in a furnace to ensure preservation of the autograph. The tile was then displayed with others on a wall, serving as a permanent souvenir of the visit. I had derived this policy following a visit I made in about 1962 to give a talk at the organic chemistry department of DuPont. There I had been asked to sign my name in colloidal gold. My collection of wall-mounted tiles at Bristol reminded me daily of many friends, and perhaps more usefully drew the attention of graduate students and others to the camaraderie that exists between scientists pursuing common goals. This positive interaction is well demonstrated by Gottfried Huttner's birthday greetings and by the letter from Helmut Werner of Würzburg reproduced on page 201.

To Professor Gordon Stone
on the occasion of his
60th birthday

Was die Chemiker suchen in Konstanz,
In Bristol schon längst Gordon Stone fands.
Der Stone, der ist gut,
Da ziehn wir den Hut
Beim Glückwunsch von außerhalb Englands.

The poem translates to

What in Konstanz the chemists are seeking,
Gordon Stone long ago found in Bristol.
That Stone, he is good,
We take off our hat
Congratulations from outside of England.

I wish you a plethora of golden stones!

Universität Konstanz

Fakultät für Chemie
Professor Dr. G. Huttner

Sincerely yours

try and in academia. Industrial interest has been sustained for many years by discoveries showing that organometallic compounds can function as catalysts in important commercial processes.[1,238] In concert, academic interest has been maintained by the sheer intellectual challenge and excitement provided by an area of chemistry that continuously produces novel results and demands new ideas. As a field of study it has attracted, and continues to attract, innovative scientists.

It is a truism to write today that advances in organometallic chemistry, as in organic chemistry, would not have been possible without the major developments in instrumentation that have occurred during the past three decades. The advent of high-field multinuclear NMR spectrometers able to operate over a range of temperatures has had a profound impact on the field, enabling chemists to study dynamic intramolecular rearrangement processes. Equally important has been the development and availability of computer-controlled automated X-ray diffraction equipment, with the associated structure-solving programs. This equipment has allowed the unraveling of unexpected structures resulting from synthetic studies and the establishment of the structures of metal cluster compounds that form an important subdivision of organometallic chemistry. Mass spectrometry and infrared spectroscopy have also played important "workhorse" roles, but have not been as influential as NMR spectroscopy or X-ray diffraction. All these physical methods for structure determination allow the acquisition of structural data at a rate undreamed of in the 1950s.

Although the days of exploratory syntheses in organometallic chemistry are by no means over, researches in this area are being increasingly directed towards applications in organic syntheses, especially metal-mediated asymmetric synthesis, providing "living" catalysts for the polymerization of block copolymers of α-alkenes, increasing our understanding of the part played by organometallic fragments in heterogeneous catalysis, and employing organometallic compounds as precursors for thin films, ceramics, and other useful materials.[239] The future of organometallic chemistry is unpredictable but is surely golden. *Comprehensive Organometallic Chemistry*[1] is currently being updated under the direction of Eddie Abel, Geoffrey Wilkinson, and myself. Not surprisingly, the most substantial volume in the revised set will be concerned with applications of organo-

Over a 16-year period at Bristol, my Labrador provided Judy and me with much exercise while at the same time the dog became well acquainted with several members of my research group (see page 85).

transition metals in organic syntheses under the editorship of L. S. Hegedus.

I was fortunate in becoming interested in organotransition metal chemistry at a time when the total knowledge of the subject could be adequately summarized in 36 pages of a small but nevertheless influential monograph.[240] There are now two primary journals[241,242] specifically concerned with new discoveries across the whole field of organometallic chemistry, a series of review volumes,[243] several excellent textbooks,[244–249] numerous specialized monographs, the nine-volume *Comprehensive Organometallic Chemistry*,[1] and a useful *Dictionary of Organometallic Compounds* published in several volumes.[250]

Increasing numbers of scientists have contributed to organometallic chemistry with a universality of effort and enthusiasm that has broken down the barriers between the classical divisions of inorganic, organic, and physical chemistry. In the development of the subject, the contributions of any one investigator and those of his co-workers must necessarily be very small. However, I believe that the work I have chosen to

My family in our Bristol garden, circa 1980. Left to right: Peter, Judith, James (holding a flower above his mother's head), and Derek Stone.

highlight in this book impinges on some of the main lines of endeavor that have developed during the past 30 or so years. To paraphrase Isaac Newton, if we have seen further it is by standing on the shoulders of giants.

Finally, a word about my family. I have mentioned earlier the unstinting support I have received for 35 years from my wife, Judith. My devotion to chemistry resulted in my spending much less time with her and with my three sons, James, Peter, and Derek, in their formative years, than I would have wished. I am proud that my sons after successful careers at university, are accomplishing much in their chosen careers in law and in medicine.

References

1. Some idea of the present wide scope of the field may be gained by examining the nine-volume work *Comprehensive Organometallic Chemistry*; Wilkinson, G.; Stone, F. G. A.; Abel, E. W., Eds.; Pergamon: Oxford, 1982, to which more than 130 experts in their respective fields contributed.

2. Wilkinson, G.; Rosenblum, M.; Whiting, M. C.; Woodward, R. B. *J. Am. Chem. Soc.* **1952**, *74*, 2125.

3. Fischer, E. O.; Pfab, W. *Z. Naturforsch., B: Anorg. Chem., Org. Chem.* **1952**, *7*, 377.

4. Fischer, E. O.; Fritz, H. P. *Adv. Inorg. Chem. Radiochem.* **1959**, *1*, 55.

5. Wilkinson, G.; Cotton, F. A. *Prog. Inorg. Chem.* **1959**, *1*, 1.

6. Fischer, E. O.; Hafner, W. *Z. Naturforsch., B: Anorg. Chem., Org. Chem.* **1955**, *10*, 655; *Z. Anorg. Allg. Chem.* **1956**, *286*, 146.

7. Hein, F. *Chem. Ber.* **1919**, *52*, 195; **1921**, *54*, 1905, 2708, 2727.

8. Zeiss, H. In *Organometallic Chemistry*; Zeiss, H., Ed.; ACS Monograph Series 147; American Chemical Society: Washington, DC, 1960; Chapter 8, pp 380–425.

9. Zeiss, H.; Tsutsui, M. *J. Am. Chem. Soc.* **1957**, *79*, 3062.

10. Hieber, W. *Adv. Organomet. Chem.* **1970**, *8*, 1.

11. Corey, E. R.; Dahl, L. F.; Beck, W. *J. Am. Chem. Soc.* **1963,** *85,* 1202.

12. Powell, H. M.; Ewens, R. V. G. *J. Chem. Soc.* **1939,** 286.

13. Cotton, F. A.; Troup, J. M. *J. Am. Chem. Soc.* **1974,** *96,* 3438.

14. Brimm, E. O.; Lynch, M. A.; Sesny, W. J. *J. Am. Chem. Soc.* **1954,** *76,* 3831.

15. A preliminary report of the structure of $[Mn_2(CO)_{10}]$ appeared in 1957; *see* Dahl, L. F.; Ishishi, E.; Rundle, R. E. *J. Chem. Phys.* **1957,** *26,* 1750. A subsequent article (Dahl, L. F.; Rundle, R. E. *Acta Crystallogr.* **1963,** *16,* 419) reported the fully refined structure.

16. Hübel, W.; Hoogzand, C. In *Organic Syntheses via Metal Carbonyls;* Wender, I.; Pino, P., Eds.; Wiley-Interscience: New York, 1968.

17. Reppe, W. *Neue Entwicklungen auf dem Gebiet der Chemie des Acetylens und Kohlenoxyds;* Springer: Berlin, 1949; pp 122–123. Reppe, W.; Schlichting, O.; Meister, H. *Justus Liebigs Ann. Chem.* **1948,** *560,* 93. Reppe, W.; Schweckendiek, W. *J. Justus Liebigs Ann. Chem.* **1948,** *560,* 104. Reppe, W.; Vetter, H. *Justus Liebigs Ann. Chem.* **1953,** *582,* 133.

18. Cotton, F. A. *Chem. Rev.* **1955,** *55,* 551.

19. Pope, W. J.; Peachey, S. J. *Proc. R. Soc. London* **1907,** *23,* 86; *J. Chem. Soc.* **1909,** *95,* 571.

20. Summers, L.; Uloth, R. H.; Holmes, A. *J. Am. Chem. Soc.* **1955,** *77,* 3604.

21. Fischer, E. O. *Angew. Chem.* **1955,** *67,* 475.

22. Wilkinson, G.; Piper, T. S. *J. Inorg. Nucl. Chem.* **1956,** *3,* 104.

23. Closson, R. D.; Kozikowski, J.; Coffield, T. H. *J. Org. Chem.* **1957,** *22,* 598.

24. Zeise, W. C. *Overs. K. Dan. Vidensk. Selsk. Forh.* **1825,** 13; *Pogg. Ann.* **1827,** *9,* 632.

25. Chatt, J. *J. Chem. Soc.* **1949,** 3340. Chatt, J.; Wilkins, R. G. *J. Chem. Soc.* **1952,** 2622.

26a. Chatt, J. *Research* **1951**, *4*, 180.

26b. The ethylidene structure was revised by Chatt after a conversation with C. K. Ingold at a meeting of the Chemical Society, now the Royal Society of Chemistry (*see* G. J. Leigh, *Coord. Chem. Rev.* **1991**, *108*, 1). Joseph Chatt reported the correct structure at a conference on cationic polymerizations organized by P. H. Plesch in March 1952, being unaware at that time of Michael Dewar's proposal for the synergistic σ and π bonding in Ag^+–alkene complexes described in *Bull. Soc. Chim. Fr.* in 1951 (*see* reference 30).

27. Remarkably, the idea of carbeneplatinum complexes appeared in reference 26a some 18 years prior to the successful identification of such species when Chatt and his co-workers prepared $[PtCl_2\{C(NHMe)(NHPh)\}(PEt_3)]$ (Bradley, E. M.; Chatt, J.; Richards, R. L.; Sim, G. A. *J. Chem. Soc., Chem. Commun.* **1969**, 1322). Moreover, Chugaev's salts, $[Pt_2(\mu\text{-}N_2H_3)_2(CNMe)_8]X_2$ (X is Cl, I, ClO_4, or N_3), first reported in 1925 (Chugaev, L.; Skanavy-Grigorieva, M.; Posniak, A. *Z. Anorg. Allg. Chem.* **1925**, *148*, 37), were shown to be mononuclear carbeneplatinum species containing the cation $[Pt\{=C(NHMe)\text{-}NHNHC=NMe\}(CNMe)_2]^+$ (Burke, A.; Balch, A. L.; Enemark, J. H. *J. Am. Chem. Soc.* **1970**, *92*, 2555). Finally, history has almost come full circle since Chatt's early proposal that Zeise's salt contained a $Pt=C(H)Me$ group, with the discovery that the μ-CHMe alkylidene fragment present in certain platinum–tungsten dimetal compounds readily rearranges to ligated ethylene (Awang, M. R.; Jeffery, J. C.; Stone, F. G. A. *J. Chem. Soc., Dalton Trans.* **1986**, 165).

28. Love, R. A.; Koetzle, T. F.; Williams, G. J. B.; Andrews, L. C.; Bau, R. *Inorg. Chem.* **1975**, *14*, 2653.

29. Chatt, J.; Duncanson, L. A. *J. Chem. Soc.* **1953**, 2939.

30. Dewar, M. J. S. *Bull. Soc. Chim. Fr.* **1951**, *18*, C79. Dewar, M. J. S.; Ford, G. P. *J. Am. Chem. Soc.* **1979**, *101*, 783.

31. Stone, F. G. A. *Chem. Rev.* **1958**, *58*, 101.

32. Seyferth, D. *Prog. Inorg. Chem.* **1962,** *3,* 129.

33. Stone, F. G. A.; Seyferth, D. *J. Inorg. Nucl. Chem.* **1955,** *1,* 112.

34. Hallam, B. F.; Pauson, P. L. *J. Chem. Soc.* **1958,** 642.

35. Reihlen, H.; Gruhl, A.; Hessling, G.; Pfrengle, O. *Justus Liebigs Ann. Chem.* **1930,** *482,* 161.

36. The useful *hapto* descriptor for defining the mode of bonding of organic fragments to metals, now in general use, was introduced much later by F. A. Cotton (*J. Am. Chem. Soc.* **1968,** *90,* 6230).

37. Bastiansen, O.; Hedberg, L.; Hedberg, K. *J. Chem. Phys.* **1957,** *27,* 1311.

38. Person, W. B.; Pimentel, G. C.; Pitzer, K. S. *J. Am. Chem. Soc.* **1952,** *74,* 3437.

39. Manuel, T. A.; Stone, F. G. A. *Proc. Chem. Soc.* **1959,** 90.

40. Manuel, T. A.; Stone, F. G. A. *J. Am. Chem. Soc.* **1960,** *82,* 366.

41. Later Pettit and co-workers (Keller, C. E.; Emerson, G. F.; Pettit, R. *J. Am. Chem. Soc.* **1965,** *87,* 1388) showed that the minor product formulated as [$Fe_2(CO)_7(C_8H_8)$] was in reality [$Fe_2(CO)_5(C_8H_8)$].

42. This compound was also reported during 1959, by two other groups (Nakamura, A.; Hagihara, N. *Bull. Chem. Soc. Jpn.* **1959,** *32,* 880. Rausch, M. D.; Schrauzer, G. N. *Chem. Ind. (London)* **1959,** 957). We had entered a period when it seemed that if one had an interesting idea for research, several others would have the same or a very similar idea at the same time.

43. Cotton, F. A. *J. Chem. Soc.* **1960,** 400.

44. Dickens, B.; Lipscomb, W. N. *J. Am. Chem. Soc.* **1961,** *83,* 4862; *J. Chem. Phys.* **1962,** *37,* 2084.

45. *Dynamic Nuclear Magnetic Resonance Spectroscopy;* Jackman, J. M.; Cotton, F. A., Eds.; Academic: New York, 1975.

46. For a complete discussion of the stereochemical nonrigidity of the molecule, *see* B. E. Mann in reference 1, volume 3, section 20.2.

47. Cotton, F. A. *Acc. Chem. Res.* **1968**, *1*, 257; *J. Organomet. Chem.* **1975**, *100*, 29. The latter article gives an entertaining and informative account of early applications of dynamic NMR spectroscopy to metal carbonyls and other organometallic species, an area where Al Cotton has made massive contributions.

48. Streitwieser, A.; Müller-Westerhoff, U. *J. Am. Chem. Soc.* **1968**, *90*, 7364. Avdeef, A.; Raymond, K. N.; Hodgson, K. O.; Zalkin, A. *Inorg. Chem.* **1972**, *11*, 1083.

49. Breil, H.; Wilke, G. *Angew. Chem., Int. Ed. Engl.* **1966**, *5*, 898.

50. Dietrich, H.; Soltwisch, M. *Angew. Chem., Int. Ed. Engl.* **1969**, *8*, 765.

51. Davidson, J. L.; Green, M.; Stone, F. G. A.; Welch, A. J. *J. Am. Chem. Soc.* **1975**, *97*, 7490; *J. Chem. Soc., Dalton Trans.* **1979**, 506.

52. Chatt, J.; Venanzi, L. M. *Nature (London)* **1956**, *177*, 852; *J. Chem. Soc.* **1957**, 4735.

53. Pettit, R. *J. Am. Chem. Soc.* **1959**, *81*, 1266.

54. Burton, R.; Green, M. L. H.; Abel, E. W.; Wilkinson, G. *Chem. Ind. (London)* **1958**, 1592.

55. Manuel, T. A.; Stone, F. G. A. *Chem. Ind. (London)* **1959**, 1349; *Chem. Ind. (London)* **1960**, 231.

56. Fischer, E. O.; Fröhlich, W. *Chem. Ber.* **1959**, *92*, 2995.

57. Bennett, M. A.; Wilkinson, G. *Chem. Ind. (London)* **1959**, 1516.

58. King, R. B.; Manuel, T. A.; Stone, F. G. A. *J. Inorg. Nucl. Chem.* **1961**, *16*, 233.

59. This complex was also reported in 1961 by Japanese workers: Nakamura, A.; Hagihara, N. *Mem. Inst. Sci. Ind. Res., Osaka Univ.* **1961**, *17*, 187.

60. Koerner von Gustorf, K.; Hogan, J. C. *Tetrahedron Lett.* **1968**, *28*, 2191.

61. Deeming, A. J.; Ullah, S. S.; Domingos, A. J. P.; Johnson, B. F. G.; Lewis, J. J. *Chem. Soc., Dalton Trans.* **1974**, 2093.

62. Arnet, J. E.; Pettit, R. *J. Am. Chem. Soc.* **1961**, *83*, 2954.

63. These include Birch, Corey, Lewis, Lillya, Pearson, and Whitesides and their respective co-workers. For reviews, see Pearson, A. J. *Acc. Chem. Res.* **1980**, *13*, 463, and also reference 1, volume 8, section 58, pp 939–1011.

64. For a personal account of Pettit's work on cyclobutadiene(tricarbonyl)iron and leading references, see *J. Organomet. Chem.* **1975**, *100*, 205. The existence of cyclobutadiene complexes of the transition metals had been predicted by H. C. Longuet-Higgins and L. E. Orgel (*J. Chem. Soc.* **1956**, 1969). Moreover, prior to the designed synthesis of **17**, two substituted cyclobutadienemetal complexes, $[Fe(CO)_3(\eta^4-C_4Ph_4)]$ (Hübel, W.; Braye, E. H. *J. Inorg. Nucl. Chem.* **1959**, *10*, 250) and $[Ni_2Cl_4(\eta^4-C_4Me_4)_2]$ (Criegee, R.; Schröder, G. *Justus Liebigs Ann. Chem.* **1959**, *623*, 1) had been reported, two discoveries no less significant because they were by chance.

65. King, R. B.; Stone, F. G. A. *J. Am. Chem. Soc.* **1960**, *82*, 3833.

66. Compound **19** had been obtained earlier from the reaction between $SnCl_2Bu^n_2$ and $K_2[Fe(CO)_4]$ but had been incorrectly formulated as the mononuclear iron compound $Bu^n_2SnFe(CO)_4$ (Hieber, W.; Breu, R. *Chem. Ber.* **1957**, *90*, 1270).

67. Cotton, J. D.; Duckworth, J. A. K.; Knox, S. A. R.; Lindley, P. F.; Paul, I.; Stone, F. G. A.; Woodward, P. *J. Chem. Soc., Chem. Commun.* **1966**, 253. Cotton, J. D.; Knox, S. A. R.; Paul, I.; Stone, F. G. A. *J. Chem. Soc. A* **1967**, 264.

68. Hoffmann, R. *Angew. Chem., Int. Ed. Engl.* **1982**, *21*, 711 (Nobel Prize lecture).

69. Abel, E. W.; Bennett, M. A.; Burton, R.; Wilkinson, G. *J. Chem. Soc.* **1958**, 4559.

70. Dauben, H. J.; Honnen, L. P. *J. Am. Chem. Soc.* **1958**, *80*, 5570.

71. King, R. B.; Stone, F. G. A. *J. Am. Chem. Soc.* **1959**, *81*, 5263.

72. Cotton, F. A.; Troup, J. M. *J. Am. Chem. Soc.* **1973**, *95*, 3798.

73. Engebretson, G.; Rundle, R. E. *J. Am. Chem. Soc.* **1963**, *85*, 481.

74. Fischer, E. O.; Ofele, K. *Chem. Ber.* **1958**, *91*, 2395.

75. Kaesz, H. D.; King, R. B.; Manuel, T. A.; Nichols, L. D.; Stone, F. G. A. *J. Am. Chem. Soc.* **1960**, *82*, 4749.

76. The reaction was also investigated by others (Green, M. L. H.; Pratt, L.; Wilkinson, G. *J. Chem. Soc.* **1960**, 989), who claimed that the product was indeed [Fe(CO)$_2$(η^5-C$_4$-H$_4$S)]. However, at the time this work was carried out the special bonding properties of the Fe(CO)$_3$ group versus Fe(CO)$_2$ fragments was not appreciated; *see* D. M. P. Mingos in reference 1, volume 3, section 19.

77. Hübel, W.; Weiss, E. *Chem. Ind. (London)* **1959**, 703. Hübel, W.; Braye, E. H. *J. Inorg. Nucl. Chem.* **1959**, *10*, 250.

78. Gates, B. C.; Katzer, J. R.; Schuit, G. C. A. *Chemistry of Catalytic Processes*; McGraw-Hill: New York, 1979; Chapter 5. Kwart, H.; Schuit, G. C. A.; Gates, B. C. *J. Catal.* **1980**, *61*, 128.

79a. King, R. B.; Treichel, P. M.; Stone, F. G. A. *J. Am. Chem. Soc.* **1961**, *83*, 3600.

79b. Holm, R. H.; King, R. B.; Stone, F. G. A. *Inorg. Chem.* **1963**, *2*, 219.

79c. Treichel, P. M.; Morris, J. H.; Stone, F. G. A. *J. Chem. Soc.* **1963**, 720.

80. Hieber, W.; Scharfenberg, C. *Chem. Ber.* **1940**, *73*, 1012. Hieber, W.; Beck, W. Z. *Anorg. Allg. Chem.* **1960**, *305*, 265.

81. King, R. B. *J. Am. Chem. Soc.* **1962**, *84*, 2460.

82. Heck, R. F.; Breslow, D. S. *J. Am. Chem. Soc.* **1960**, *82*, 750.

83. Prichard, W. W. U.S. Patent 2 600 571, 1952. *Chem. Abstr.* **1952,** *46,* P10188f.

84a. Smidt, J.; Hafner, W. *Angew. Chem.* **1959,** *71,* 284.

84b. Shaw, B. L.; Sheppard, N. *Chem. Ind. (London)* **1961,** 517. Shaw, B. L. *Chem. Ind. (London)* **1962,** 1190.

85. Kaesz, H. D.; King, R. B.; Stone, F. G. A. *Z. Naturforsch., B: Anorg. Chem., Org. Chem.* **1960,** *15,* 682.

86. McClellan, W. R.; Hoehn, H. H.; Cripps, H. N.; Muetterties, E. L.; Howk, B. W. *J. Am. Chem. Soc.* **1961,** *83,* 1601.

87. Chatt, J.; Shaw, B. L. *J. Chem. Soc.* **1959,** 705, 4020; **1960,** 1718; **1961,** 285.

88. Piper, T. S.; Wilkinson, G. *Naturwissenschaften* **1956,** *43,* 129.

89. Emeléus, H. J. *Angew. Chem., Int. Ed. Engl.* **1962,** *1,* 129; *J. Chem. Soc.* **1954,** 2979.

90. Banks, R. E.; Haszeldine, R. N. *Adv. Inorg. Chem. Radiochem.* **1961,** *3,* 337. Lagowski, J. J. *Q. Rev.* **1959,** *13,* 233.

91. Kaesz, H. D.; King, R. B.; Stone, F. G. A. *Z. Naturforsch., B: Anorg. Chem., Org. Chem.* **1960,** *15,* 763.

92. Coffield, T. H.; Kozikowski, J.; Closson, R. D. *Abstracts of the I. C. C. Conference;* The Chemical Society: London, 1959; p 126; Special Publication No. 13.

93. McClellan, W. R. *J. Am. Chem. Soc.* **1961,** *83,* 1598.

94. For a review and a listing of primary journal articles, *see* Treichel, P. M.; Stone, F. G. A. *Adv. Organomet. Chem.* **1964,** *1,* 143.

95. Pitcher, E.; Buckingham, A. D.; Stone, F. G. A. *J. Chem. Phys.* **1962,** *36,* 124.

96. The contents of this course were subsequently to appear in the influential book by F. A. Cotton, *Chemical Applications of Group Theory;* Interscience: New York, 1963, now in its third edition and translated into six languages, but not including Texan.

97. Coyle, T. D.; Stone, F. G. A. *J. Chem. Phys.* **1960,** *32,* 1892.

98. For a review, *see* Davidson, P. J.; Lappert, M. F.; Pearce, R. *Chem. Rev.* **1976,** *76,* 219. For examples of specific compounds, *see* the various sections of reference 1.

99. Wilke, G. *Pure Appl. Chem.* **1978,** *50,* 677.

100. King, R. B.; Treichel, P. M.; Stone, F. G. A. *J. Am. Chem. Soc.* **1961,** *83,* 3593. Coyle, T. D.; King, R. B.; Pitcher, E.; Stafford, S. L.; Treichel, P. M.; Stone, F. G. A. *J. Inorg. Nucl. Chem.* **1961,** *20,* 172.

101. Collman, J. P.; Roper, W. R. *Adv. Organomet. Chem.* **1968,** *7,* 53.

102. Halpern, J. *Chem. Eng. News.* **1966,** *Oct. 31,* 68–75. Halpern, J. In *Homogeneous Catalysis;* Luberoff, B. J., Ed.; Advances in Chemistry 70; American Chemical Society: Washington, DC, 1968; pp 1–24; *Faraday Discuss. Chem. Soc.* **1969,** *46,* 7.

103. J. Chatt and B. L. Shaw (*J. Chem. Soc.* **1962,** 5075) reported a reversible reaction between [PtH(Cl)(PEt$_3$)$_2$] and C$_2$H$_4$ to give [PtCl(Et)(PEt$_3$)$_2$] and would thus appear to have been the first to demonstrate the important process L$_n$M–H + CH$_2$=CH$_2$ → L$_n$M–CH$_2$CH$_3$.

104. Treichel, P. M.; Chaudhari, M. A.; Stone, F. G. A. *J. Organomet. Chem.* **1963,** *1,* 98; **1964,** *2,* 206. Phillips, J. R.; Rosevear, D. T.; Stone, F. G. A. *J. Organomet. Chem.* **1964,** *2,* 455. Jolly, P. W.; Stone, F. G. A. *J. Chem. Soc., Chem. Commun.* **1965,** 85.

105. Bruce, M. I.; Stone, F. G. A. *Prep. Inorg. React.* **1968,** *4,* 177.

106. Usón, R. *Pure Appl. Chem.* **1986,** *58,* 647. Usón, R.; Laguna, A. *Coord. Chem. Rev.* **1986,** *70,* 1. Usón, R.; Forniés, J. *Adv. Organomet. Chem.* **1988,** *28,* 219.

107. Cramer, R.; Parshall, G. W. *J. Am. Chem. Soc.* **1965,** *87,* 1392. Parshall, G. W.; Jones, F. N. *J. Am. Chem. Soc.* **1965,** *87,* 5356.

108. Cramer, R.; Kline, J. B.; Roberts, J. D. *J. Am. Chem. Soc.* **1969,** *91,* 2519.

109. Malatesta, L.; Cariello, C. *J. Chem. Soc.* **1958,** 2323.

110. Green, M.; Osborn, R. B. L.; Rest, A. J.; Stone, F. G. A. *J.*

Chem. Soc., Chem. Commun. **1966,** 502; *J. Chem. Soc. A* **1968,** 2525. Maples, P. K.; Green, M.; Stone, F. G. A. *J. Chem. Soc., Dalton Trans.* **1973,** 2069.

111. Wilke, G. *Angew. Chem.* **1960,** *72,* 581. Bogdanovic, B.; Kroner, M.; Wilke, G. *Justus Liebigs Ann. Chem.* **1966,** *699,* 1.

112. Stone, F. G. A. *Pure Appl. Chem.* **1972,** *30,* 551.

113. Mond, L.; Hirtz, N.; Cowap, M. D. *Proc. Chem. Soc.* **1910,** *26,* 67; *J. Chem. Soc.* **1910,** 97, 798; *Z. Anorg. Allg. Chem.* **1910,** *68,* 207.

114. Corey, E. R.; Dahl, L. F. *J. Am. Chem. Soc.* **1961,** *83,* 2203.

115. Mason, R.; Rae, A. I. M. *J. Chem. Soc. A* **1968,** 778.

116. Churchill, M. R.; Hollander, F. J.; Hutchinson, J. P. *Inorg. Chem.* **1977,** *16,* 2655.

117. Bruce, M. I.; Stone, F. G. A. *J. Chem. Soc., Chem. Commun.* **1966,** 684; *J. Chem. Soc. A* **1967,** 1238.

118. For a tabulated survey of eight alternative syntheses, *see* M. I. Bruce in reference 1, Volume 4, Section 32.2. A simple synthesis has been described by Bruce, M. I.; Jensen, C. M.; Jones, N. L. *Inorg. Synth.* **1990,** *28,* 216.

119. Bruce, M. I.; Stone, F. G. A. *Angew. Chem., Int. Ed. Engl.* **1968,** *7,* 427.

120. King, R. B.; Stone, F. G. A. *Inorg. Synth.* **1963,** *7,* 193.

121. *See* Shriver, D. F.; Whitmire, K. H. In reference 1, Volume 4, Section 31.1, and Davidson, J. L. In reference 1, Volume 4, Section 31.5.

122. Bruce, M. I. *J. Organomet. Chem.* **1990,** *400,* 321.

123. Cotton, J. D.; Bruce, M. I.; Stone, F. G. A. *J. Chem. Soc. A* **1968,** 2162.

124. Yawney, D. B. W.; Stone, F. G. A. *J. Chem. Soc., Chem. Commun.* **1968,** 619; *J. Chem. Soc. A* **1969,** 502.

125. Roberts, D. A.; Geoffroy, G. L. In reference 1, Volume 6, Section 40.

126. Cotton, J. D.; Knox, S. A. R.; Stone, F. G. A. *J. Chem. Soc., Chem. Commun.* **1967**, 965; *J. Chem. Soc. A* **1968**, 2758. Knox, S. A. R.; Stone, F. G. A. *J. Chem. Soc. A* **1969**, 2559; **1971**, 2874. Brookes, A.; Knox, S. A. R.; Stone, F. G. A. *J. Chem. Soc. A* **1971**, 3469.

127. Osmium analogs of **54–57** were also prepared in our laboratory, but the reader's attention is drawn to the studies of Bill Graham and his co-workers, who carried out comprehensive research on the molecules $[M(ER_3)_2(CO)_4]$ (M is Fe, Ru, or Os; E is Si, Ge, Sn, or Pb; R is halogen, Me, and so forth). This work, involving detailed IR and NMR (^1H and ^{13}C-$\{^1H\}$) measurements, showed that these complexes provide excellent examples of stereospecific intramolecular ligand-exchange processes and fluxional behavior. *See* Pomeroy, R. K.; Vancea, L.; Calhoun, H. P.; Graham, W. A. G. *Inorg. Chem.* **1977**, *16*, 1508 and references cited therein.

128. Brookes, A.; Howard, J. A. K.; Knox, S. A. R.; Stone, F. G. A.; Woodward, P. *J. Chem. Soc., Chem. Commun.* **1973**, 587.

129. Landesberg, J. M. In *The Organic Chemistry of Iron*; Koerner von Gustorf, E. A.; Grevels, F.-W.; Fischler, I., Eds.; Academic: New York, 1978; Volume 1, Chapter 12.

130a. Harris, P. J.; Howard, J. A. K.; Knox, S. A. R.; McKinney, R. J.; Phillips, R. P.; Stone, F. G. A.; Woodward, P. *J. Chem. Soc., Dalton Trans.* **1978**, 403.

130b. Knox, S. A. R.; McKinney, R. J.; Riera, V.; Stone, F. G. A.; Szary, A. C. *J. Chem. Soc., Dalton Trans.* **1979**, 1801.

130c. Howard, J. A. K.; Stansfield, R. F. D.; Woodward, P. *J. Chem. Soc., Dalton Trans.* **1979**, 1812.

130d. Knox, S. A. R.; McKinney, R. J.; Stone, F. G. A. *J. Chem. Soc., Dalton Trans.* **1980**, 235.

130e. Howard, J. A. K.; Knox, S. A. R.; Riera, V.; Stone, F. G. A.; Woodward, P. *J. Chem. Soc., Chem. Commun.* **1974**, 452.

131. The reaction between cyclooctatetraene and $[Ru_3(CO)_{12}]$ is complex, leading to at least eight ruthenium complexes.

For a detailed review, *see* M. I. Bruce in reference 1, Volume 4, Section 32.5.

132. Knox, S. A. R.; Stone, F. G. A. *Acc. Chem. Res.* **1974**, *7*, 321.

133. Hawthorne, M. F.; Young, D. C.; Wegner, P. A. *J. Am. Chem. Soc.* **1965**, *87*, 1818. Callahan, K. P.; Hawthorne, M. F. *Adv. Organomet. Chem.* **1976**, *14*, 145.

134. Spencer, J. L.; Green, M.; Stone, F. G. A. *J. Chem. Soc., Chem. Commun.* **1972**, 1178.

135. Smith, H. D.; Hawthorne, M. F. *Inorg. Chem.* **1974**, *13*, 2312. Manning, M. J.; Knobler, C. B.; Hawthorne, M. F. *J. Am. Chem. Soc.* **1988**, *110*, 4458. Manning, M. J.; Knobler, C. B.; Hawthorne, M. F. *Inorg. Chem.* **1991**, *30*, 3589. Kim, J.; Do, Y.; Sohn, Y. S.; Knobler, C. B.; Hawthorne, M. F. *J. Organomet. Chem.* **1991**, *418*, C1.

136. Colquhoun, H. M.; Greenhough, T. J.; Wallbridge, M. G. H. *J. Chem. Soc., Chem. Commun.* **1977**, 737.

137. Dunks, G. B.; Hawthorne, M. F. *J. Am. Chem. Soc.* **1970**, *92*, 7213.

138. Dunks, G. B.; McKown, M. M.; Hawthorne, M. F. *J. Am. Chem. Soc.* **1971**, *93*, 2541.

139. Gerlach, D. H.; Kane, A. R.; Parshall, G. W.; Jesson, J. P.; Muetterties, E. L. *J. Am. Chem. Soc.* **1971**, *93*, 3543.

140. Green, M.; Spencer, J. L.; Stone, F. G. A. *J. Chem. Soc., Dalton Trans.* **1979**, 1679 and references therein.

141. Stone, F. G. A. *J. Organomet. Chem.* **1975**, *100*, 257.

142. Müller, J.; Göser, P. *Angew. Chem., Int. Ed. Engl.* **1967**, *6*, 364.

143. Jolly, P. W.; Wilke, G. *The Organic Chemistry of Nickel*; Academic: New York, 1974; Volumes 1 and 2.

144. Green, M.; Howard, J. A. K.; Spencer, J. L.; Stone, F. G. A. *J. Chem. Soc., Chem. Commun.* **1975**, 3; *J. Chem. Soc., Dalton Trans.* **1977**, 271.

145. Spencer, J. L. *Inorg. Synth.* **1979**, *19*, 213. Crascall, L. E.; Spencer, J. L. *Inorg. Synth.* **1990**, *28*, 126.

146. Boag, N. M.; Howard, J. A. K.; Spencer, J. L.; Stone, F. G. A. *J. Chem. Soc., Dalton Trans.* **1981**, 1051.

147. Green, M.; Howard, J. A. K.; Murray, M.; Spencer, J. L.; Stone, F. G. A. *J. Chem. Soc., Dalton Trans.* **1977**, 1509.

148. Fischer, K.; Jonas, K.; Wilke, G. *Angew. Chem., Int. Ed. Engl.* **1973**, *12*, 565.

149. Stone, F. G. A. *Acc. Chem. Res.* **1981**, *14*, 318.

150. Howard, J. A. K.; Spencer, J. L.; Mason, S. A. *Proc. R. Soc. London A* **1983**, *386*, 161.

151. Green, M.; Howard, J. A. K.; Spencer, J. L.; Stone, F. G. A. *J. Chem. Soc., Chem. Commun.* **1975**, 449.

152. Rösch, N.; Hoffmann, R. *Inorg. Chem.* **1974**, *13*, 2656.

153. Barker, G. K.; Green, M.; Howard, J. A. K.; Spencer, J. L.; Stone, F. G. A. *J. Chem. Soc., Dalton Trans.* **1978**, 1839.

154. Green, M.; Scholes, G.; Stone, F. G. A. *J. Chem. Soc., Dalton Trans.* **1978**, 309.

155. Boag, N. M.; Green, M.; Howard, J. A. K.; Stone, F. G. A.; Wadepohl, H. *J. Chem. Soc., Dalton Trans.* **1981**, 862 and references therein.

156. Ciriano, M.; Green, M.; Howard, J. A. K.; Proud, J.; Spencer, J. L.; Stone, F. G. A. *J. Chem. Soc., Dalton Trans.* **1978**, 801. Auburn, M.; Ciriano, M.; Howard, J. A. K.; Murray, M.; Pugh, N. J.; Spencer, J. L.; Stone, F. G. A. *J. Chem. Soc., Dalton Trans.* **1980**, 659.

157. For leading references *see* Graham, W. A. G. *J. Organomet. Chem.* **1986**, *300*, 81.

158. Green, M.; Spencer, J. L.; Stone, F. G. A.; Tsipis, C. A. *J. Chem. Soc., Dalton Trans.* **1977**, 1519, 1525. Barlow, A. P.; Boag, N. M.; Stone, F. G. A. *J. Organomet. Chem.* **1980**, *191*, 39.

159. Green, M.; Laguna, A.; Spencer, J. L.; Stone, F. G. A. *J. Chem. Soc., Dalton Trans.* **1977**, 1010.

160. Cotton, F. A.; Walton, R. A. *Multiple Bonds Between Metal Atoms*, 2nd ed.; Oxford University Press: Oxford, UK, 1993.

161. *Transition Metal Clusters*; Johnson, B. F. G., Ed.; Wiley–Interscience: New York, 1980. Mingos, D. M. P.; Wales, D. J. *Introduction to Cluster Chemistry*; Prentice-Hall: Englewood Cliffs, NJ, 1990. *The Chemistry of Metal Cluster Complexes*; Shriver, D. F.; Kaesz, H. D.; Adams, R. D., Eds.; VCH: Weinheim, Germany, 1990.

162. Chini, P.; Colli, L.; Peraldo, M. *Gazz. Chim. Ital.* **1960**, *90*, 1005.

163. Abel, E. W.; Singh, A.; Wilkinson, G. W. *J. Chem. Soc.* **1960**, 1321.

164. Tilney-Bassett, J. F. *Proc. Chem. Soc.* **1960**, 419.

165. King, R. B.; Treichel, P. M.; Stone, F. G. A. *Chem. Ind. (London)* **1961**, 747.

166. Joshi, K. K.; Pauson, P. L. *Z. Naturforsch., B: Anorg. Chem., Org. Chem.* **1962**, *17*, 565.

167. Compounds with bonds between metal carbonyl groups and some of the post transition elements (e.g., Ag, Zn, or Hg) had been reported in the scientific literature.

168. Barker, G. K.; Garcia, M. P.; Green, M.; Stone, F. G. A.; Welch, A. J. *J. Chem. Soc., Dalton Trans.* **1982**, 1679 and references therein.

169. Johnson, B. F. G.; Lewis, J.; Kilty, P. A. *J. Chem. Soc. A* **1968**, 2859.

170. Farrugia, L. J.; Howard, J. A. K.; Mitrprachachon, P.; Spencer, J. L.; Stone, F. G. A.; Woodward, P. *J. Chem. Soc., Chem. Commun.* **1978**, 260.

171. Farrugia, L. J.; Howard, J. A. K.; Mitrprachachon, P.; Stone, F. G. A.; Woodward, P. *J. Chem. Soc., Dalton Trans.* **1981**, 155, 162, 171.

172. Stone, F. G. A. *Inorg. Chim. Acta* **1981**, *50*, 33; *Proc. R. Soc. London A* **1982**, *308*, 87.

173. Farrugia, L. J. *Adv. Organomet. Chem.* **1990**, *31*, 301.

174. Farrugia, L. J.; Green, M.; Hankey, D. R.; Murray, M.; Orpen, A. G.; Stone, F. G. A. *J. Chem. Soc., Dalton Trans.* **1985**, 177.

175. Adams, R. D.; Horváth, I. T. *Prog. Inorg. Chem.* **1985**, *33*, 127.

176. Davies, D. L.; Jeffery, J. C.; Miguel, D.; Stone, F. G. A. *J. Organomet. Chem.* **1990**, *383*, 463.

177. Pinhas, A. R.; Albright, T. A.; Hofmann, P.; Hoffmann, R. *Helv. Chim. Acta* **1980**, *63*, 29.

178. Nutton, A.; Maitlis, P. M. *J. Organomet. Chem.* **1979**, *166*, C21.

179. Green, M.; Mills, R. M.; Pain, G. N.; Stone, F. G. A.; Woodward, P. *J. Chem. Soc., Dalton Trans.* **1982**, 1309. Aldridge, M. L.; Green, M.; Howard, J. A. K.; Pain, G. N.; Porter, S. J.; Stone, F. G. A.; Woodward, P. *J. Chem. Soc., Dalton Trans.* **1982**, 1333. Barr, R. D.; Green, M.; Marsden, K.; Stone, F. G. A.; Woodward, P. *J. Chem. Soc., Dalton Trans.* **1983**, 507.

180. Green, M.; Howard, J. A. K.; Pain, G. N.; Stone, F. G. A. *J. Chem. Soc., Dalton Trans.* **1982**, 1327.

181. Boag, N. M.; Green, M.; Mills, R. M.; Pain, G. N.; Stone, F. G. A.; Woodward, P. *J. Chem. Soc., Chem. Commun.* **1980**, 1171.

182. Cirjak, L. M.; Huang, J.-S.; Zhu, Z.-H.; Dahl, L. F. *J. Am. Chem. Soc.* **1980**, *102*, 6623.

183. Barr, R. D.; Green, M.; Howard, J. A. K.; Marder, T. B.; Orpen, A. G.; Stone, F. G. A. *J. Chem. Soc., Dalton Trans.* **1984**, 2757.

184. Fischer, E. O. *Adv. Organomet. Chem.* **1976**, *14*, 1; *Angew. Chem.* **1974**, *86*, 651 (Nobel Prize lecture).

185. Fischer, E. O.; Maasböl, A. *Angew. Chem., Int. Ed. Engl.* **1964**, *3*, 580.

186. Fischer, E. O.; Kreis, G.; Kreiter, C. G.; Müller, J.; Huttner, G.; Lorenz, H. *Angew. Chem., Int. Ed. Engl.* **1973**, *12*, 564.

187. Schrock, R. R. *Acc. Chem. Res.* **1979**, *12*, 98; *J. Organomet. Chem.* **1986**, *300*, 249.

188. Cardin, D. J.; Cetinkaya, B.; Lappert, M. F. *Chem. Rev.* **1972**,

72, 545. *Advances in Metal Carbene Chemistry*; Schubert, U., Ed.; Kluwer Academic: Dordrecht, Netherlands, 1989.

189. Gallop, M. A.; Roper, W. R. *Adv. Organomet. Chem.* **1986**, *25*, 121. Kim, H. P.; Angelici, R. J. *Adv. Organomet. Chem.* **1987**, *27*, 51.

190. Mayr, A.; Hoffmeister, H. *Adv. Organomet. Chem.* **1991**, *32*, 227.

191. Stone, F. G. A. *Angew. Chem., Int. Ed. Engl.* **1984**, *23*, 89; *Pure Appl. Chem.* **1986**, *58*, 529.

192. Markby, R.; Wender, I.; Friedel, R. A.; Cotton, F. A.; Sternberg, H. W. *J. Am. Chem. Soc.* **1958**, *80*, 6529.

193. Sutton, P. W.; Dahl, L. F. *J. Am. Chem. Soc.* **1967**, *89*, 261.

194. Seyferth, D. *Adv. Organomet. Chem.* **1976**, *14*, 97.

195. Fischer, E. O.; Beck, H.-J. *Angew. Chem., Int. Ed. Engl.* **1970**, *9*, 72.

196. Herrmann, W. A.; Reiter, B.; Biersack, H. *J. Organomet. Chem.* **1975**, *97*, 245. Herrmann, W. A.; Krüger, C.; Goddard, R.; Bernal, I. *J. Organomet. Chem.* **1977**, *140*, 73.

197. Green, M.; Howard, J. A. K.; Laguna, A.; Murray, M.; Spencer, J. L.; Stone, F. G. A. *J. Chem. Soc., Chem. Commun.* **1975**, 451.

198. Ashworth, T. V.; Howard, J. A. K.; Stone, F. G. A. *J. Chem. Soc., Chem. Commun.* **1979**, 42.

199. Sumner, C. E.; Riley, P. E.; Davis, R. E.; Pettit, R. *J. Am. Chem. Soc.* **1980**, *102*, 1752. Sumner, C. E.; Collier, J. A.; Pettit, R. *Organometallics* **1982**, *1*, 1350 and references therein.

200. Muetterties, E. L.; Stein, J. *Chem. Rev.* **1979**, *79*, 479. Tachikawa, M.; Muetterties, E. L. *Prog. Inorg. Chem.* **1981**, *28*, 203.

201. Herrmann, W. A. *Adv. Organomet. Chem.* **1982**, *20*, 159; *Angew. Chem., Int. Ed. Engl.* **1982**, *21*, 117.

202. *Organometallic Compounds: Synthesis, Structure, and Theory*; Shapiro, B. L., Ed.; Texas A&M University Press: College Station, TX, 1983.

203. Moskovits, M. *Acc. Chem. Res.* **1979**, *12*, 229.

204a. Bergman, R. G.; Theopold, K. H. *Organometallics* **1982**, *1*, 219, 1571. Seidler, P. F.; Bryndza, H. E.; Frommer, J. E.; Stuhl, L. S.; Bergman, R. G. *Organometallics* **1983**, *2*, 1701.

204b. Jacobsen, E. N.; Goldberg, K. I.; Bergman, R. G. *J. Am. Chem. Soc.* **1988**, *110*, 3706. Hostetler, M. J.; Bergman, R. G. *J. Am. Chem. Soc.* **1992**, *114*, 762. Hostetler, M. J.; Butts, M. D.; Bergman, R. G. *Organometallics* **1993**, *12*, 65.

205. Casey, C. P.; Meszaros, M. W.; Fagan, P. J.; Bly, R. K.; Marder, S. R.; Austin, E. A. *J. Am. Chem. Soc.* **1986**, *108*, 4043. Casey, C. P.; Gable, K. P.; Roddick, D. M. *Organometallics* **1990**, *9*, 221, and references cited therein.

206. Chisholm, M. H.; Heppert, J. A. *Adv. Organomet. Chem.* **1986**, *26*, 97. Buhro, W. E.; Chisholm, M. H. *Adv. Organomet. Chem.* **1987**, *27*, 311. Chisholm, M. H. *J. Organomet. Chem.* **1990**, *400*, 235.

207. Knox, S. A. R. *Pure Appl. Chem.* **1984**, *56*, 81; *J. Organomet. Chem.* **1990**, *400*, 255.

208. Anderson, O. P.; Bender, B. R.; Norton, J. R.; Larson, A. C.; Vergamini, P. J. *Organometallics* **1991**, *10*, 3145, and references cited therein.

209. Laws, W. J.; Puddephatt, R. J. *J. Chem. Soc., Chem. Commun.* **1984**, 116. Puddephatt, R. J. *Polyhedron* **1988**, *7*, 767.

210. Shultz, A. J.; Williams, J. M.; Calvert, R. B.; Shapley, J. R.; Stucky, G. D. *Inorg. Chem.* **1979**, *18*, 319. Shapley, J. R.; Cree-Uchiyama, M. E.; St. George, G. M.; Churchill, M. R.; Bueno, C. *J. Am. Chem. Soc.* **1983**, *105*, 140. Chi, Y.; Shapley, J. R. *Organometallics* **1985**, *4*, 1900. Cree-Uchiyama, M.; Shapley, J. R.; St. George, G. M. *J. Am. Chem. Soc.* **1986**, *108*, 1316.

211. Ashworth, T. V.; Berry, M.; Howard, J. A. K.; Laguna, M.; Stone, F. G. A. *J. Chem. Soc., Chem. Commun.* **1979**, 43, 45.

212. For leading references to the heteronuclear dimetal compounds with bridging carbene ligands, *see* Howard, J. A. K.; Mead, K. A.; Moss, J. R.; Navarro, R.; Stone, F. G. A.; Woodward, P. J. *Chem. Soc., Dalton Trans.* **1981**,

743. Jeffery, J. C.; Moore, I.; Murray, M.; Stone, F. G. A. *J. Chem. Soc., Dalton Trans.* **1982,** 1741.

213. Chatt, J.; Rowe, G. A.; Williams, A. A. *Proc. Chem. Soc.* **1957,** 208.

214. Stone, F. G. A. In *Inorganic Chemistry: Toward the 21st Century*; Chisholm, M. H., Ed.; ACS Symposium Series 211; American Chemical Society: Washington, DC, 1983; pp 383–395.

215. Garcia, M. E.; Jeffery, J. C.; Sherwood, P.; Stone, F. G. A. *J. Chem. Soc., Chem. Commun.* **1986,** 802. Byrne, P. G.; Garcia, M. E.; Jeffery, J. C.; Sherwood, P.; Stone, F. G. A. *J. Chem. Soc., Chem. Commun.* **1987,** 53.

216. Delgado, E.; Farrugia, L. J.; Hein, J.; Jeffery, J. C.; Ratermann, A. L.; Stone, F. G. A. *J. Chem. Soc., Dalton Trans.* **1987,** 1191.

217. Jeffery, J. C.; Went, M. J. *Polyhedron* **1988,** 7, 775.

218. Chetcuti, M. J.; Chetcuti, P. A. M.; Jeffery, J. C.; Mills, R. M.; Mitrprachachon, P.; Pickering, S. J.; Stone, F. G. A.; Woodward, P. *J. Chem. Soc., Dalton Trans.* **1982,** 699.

219. Green, M.; Porter, S. J.; Stone, F. G. A. *J. Chem. Soc., Dalton Trans.* **1983,** 513.

220. Green, M.; Jeffery, J. C.; Porter, S. J.; Razay, H.; Stone, F. G. A. *J. Chem. Soc., Dalton Trans.* **1982,** 2475.

221. Carriedo, G. A.; Howard, J. A. K.; Stone, F. G. A. *J. Chem. Soc., Dalton Trans.* **1984,** 1555.

222. Ashworth, T. V.; Chetcuti, M. J.; Howard, J. A. K.; Stone, F. G. A.; Wisbey, S. J.; Woodward, P. *J. Chem. Soc., Dalton Trans.* **1981,** 763. Davies, S. J.; Hill, A. F.; Pilotti, M. U.; Stone, F. G. A. *Polyhedron* **1989,** 8, 2265.

223. Elliott, G. P.; Howard, J. A. K.; Mise, T.; Moore, I.; Nunn, C. M.; Stone, F. G. A. *J. Chem. Soc., Dalton Trans.* **1986,** 2091. Elliott, G. P.; Howard, J. A. K.; Nunn, C. M.; Stone, F. G. A. *J. Chem. Soc., Chem. Commun.* **1986,** 431.

224a. Elliott, G. P.; Howard, J. A. K.; Mise, T.; Nunn, C. M.; Stone, F. G. A. *Angew. Chem., Int. Ed. Engl.* **1986,** 25, 190.

224b. Elliott, G. P.; Howard, J. A. K.; Mise, T.; Nunn, C. M.; Stone, F. G. A. *J. Chem. Soc., Dalton Trans.* **1987,** 2189.

225a. Davies, S. J.; Stone, F. G. A. *J. Chem. Soc., Dalton Trans.* **1989,** 785.

225b. Davies, S. J.; Howard, J. A. K.; Musgrove, R. J.; Stone, F. G. A. *J. Chem. Soc., Dalton Trans.* **1989,** 2269.

226. The connection between teaching and research in universities has been ably defended by J. H. Horlock in *Sci. Publ. Affairs* **1991,** *6,* 77 published by the Royal Society. Dr. Horlock, an engineer by training, was head of Britain's Open University for several years.

227. University Grants Committee, report of the chemistry review; Stone, F. G. A., Chairman *University Chemistry— The Way Forward;* H.M.S.O.: London, 1988.

228. Lord Porter. Presidential Anniversary Address to the Royal Society; *Sci. Publ. Affairs* **1991,** *6,* 3.

229. *Opportunities in Chemistry;* National Academy: Washington, DC, 1985. Pimentel, G. C.; Coonrod, J. A. *Opportunities in Chemistry, Today and Tomorrow;* National Academy: Washington, DC, 1987.

230. Right Honorable Margaret Thatcher F. R. S.; Speech at the Dinner of the Royal Society, September 27th, 1988, published in the *Royal Society News* **1988,** 4(11).

231. Stone, F. G. A. *Adv. Organomet. Chem.* **1990,** *31,* 53. Brew, S. A.; Stone, F. G. A. *Adv. Organomet. Chem.* **1993,** *35,* 135.

232. Curtis, L. A.; Pople, J. A. *J. Chem. Phys.* **1988,** *88,* 7405.

233. Berkowitz, J.; Mayhew, C. A.; Ruščić, B. *J. Chem. Phys.* **1988,** *88,* 7396.

234. Bartocha, B.; Stone, F. G. A. *Z. Naturforsch., B: Anorg. Chem., Org. Chem.* **1958,** *13,* 347.

235. Stafford, S. L.; Stone, F. G. A. *J. Am. Chem. Soc.* **1960,** *82,* 6238.

236. Massey, A. G.; Park, A. J.; Stone, F. G. A. *Proc. Chem. Soc.* **1963,** 212. Massey, A. G.; Park, A. J. *J. Organomet. Chem.* **1964,** *2,* 245; **1966,** *5,* 218.

237. Yang, X.; Stern, C. L.; Marks, T. J. *J. Am. Chem. Soc.* **1991,** *113,* 3623.

238. Parshall, G. W.; Ittel, S. D. *Homogeneous Catalysis,* 2nd ed.; Wiley: New York, 1993.

239. Green, M. L. H. *J. Chem. Soc., Dalton Trans.* **1991,** 575.

240. Coates, G. E. *Organo-Metallic Compounds;* Methuen: London, 1956.

241. *J. Organomet. Chem.* Elsevier Science Publishing Co., Lausanne, Switzerland.

242. *Organometallics,* published by the American Chemical Society.

243. *Advances in Organometallic Chemistry;* Stone, F. G. A.; West, R., Eds.; Academic: San Diego, CA; Vols. 1–35.

244. Yamamoto, A. *Organotransition Metal Chemistry;* Wiley–Interscience: New York, 1986.

245. Collman, J. P.; Hegedus, L. S.; Norton, J. R.; Finke, R. G. *Principles and Applications of Organotransition Metal Chemistry,* 2nd ed.; University Science Books: Mill Valley, CA, 1987.

246. Lukehart, C. M. *Fundamental Transition Metal Organometallic Chemistry;* Brooks-Cole: Monterey, CA, 1985.

247. Elschenbroich, Ch.; Salzer, A. *Organometallics,* 2nd ed.; VCH: Weinheim, Germany, 1992.

248. Crabtree, R. H. *The Organometallic Chemistry of the Transition Metals;* Wiley: New York, 1988.

249. Pearson, A. J. *Metallo-organic Chemistry;* Wiley–Interscience: New York, 1985.

250. *Dictionary of Organometallic Compounds;* Chapman and Hall: London, 1984; published in 3 volumes with 5 supplementary volumes and 2 index volumes.

251. Dossett, S. J.; Hill, A. F.; Jeffery, J. C.; Marken, F.; Sherwood, P.; Stone, F. G. A. *J. Chem. Soc., Dalton Trans.* **1988,** 2453. Brew, S. A.; Dossett, S. J.; Jeffery, J. C.; Stone, F. G. A. *J. Chem. Soc., Dalton Trans.* **1990,** 3709.

Index

A

Copy editing and indexing: Steven Powell
Production: Paula M. Bérard

Production Manager: Robin Giroux

Printed and bound by Maple Press, York, PA